EXS 91

Genes and Mechanisms in Vertebrate Sex Determination

Edited by G. Scherer and M. Schmid

Birkhäuser Verlag
Basel · Boston · Berlin

Editors

PD Dr. Gerd Scherer
Institute of Human Genetics
and Anthropology
University of Freiburg
Breisacher Strasse 33
79106 Freiburg
Germany

Prof. Dr. Michael Schmid
Department of Human Genetics
University of Würzburg
Biozentrum
Am Hubland
97074 Würzburg
Germany

The reviews collected in this monograph are updated and revised versions of the articles published in the multi-author review on "Genes and mechanisms in vertebrate sex determination" in Cellular and Molecular Life Sciences 55: 821–931 (1999), except for the unchanged review by S. Ohno and the new review by M. Schmid and C. Steinlein.

Library of Congress Cataloging-in-Publication Data
Genes and mechanisms in vertebrate sex determination / edited by G. Scherer and M. Schmid.
    p. cm. -- (EXS; 91)
  Includes bibliographical references and index.
  ISBN 3764361689 (alk. paper)
    1. Sex determination, Genetic. I. Scherer, G. (Gerd), 1948- II. Schmid, M. Michael.
  III. Series.

QP278.5 .G46 2001
571.8′8216--dc21   00-048566

Deutsche Bibliothek Cataloging-in-Publication Data
Genes and mechanisms in vertebrate sex determination / ed. by G. Scherer and M. Schmid. –
Basel ; Boston ; Berlin : Birkhäuser, 2001
  (EXS; 91)
  ISBN 3-7643-6168-9

ISBN 3-7643-6168-9 Birkhäuser Verlag, Basel – Boston – Berlin

© 2001 Birkhäuser Verlag, P.O. Box 133, CH-4010 Basel, Switzerland
Printed on acid-free paper produced from chlorine-free pulp. TCF ∞
Cover illustration: Julius Bissier, männlich-weibliches Einheitssymbol, 1934. © VG Bild-Kunst, Bonn 2000
Printed in Germany
ISBN 3-7643-6168-9
9 8 7 6 5 4 3 2 1

The authors dedicate this book to the memory of Susumu Ohno, whose contributions to the field of vertebrate sex determination have stimulated and inspired research in this field over many years.

Contents

List of contributors

Jean-François Baroiller, CIRAD-EMVT (Centre International en Recherche
Agronomique pour le Développement), Laboratoire de Physiologie
des Poissons, Campus de Beaulieu, 35042 Rennes Cedex, France;
e-mail: jfb@beaulieu.rennes.inra.fr

Giovanna Camerino, Biologia Generale e Genetica Medica,
Università di Pavia, via Forlanini 14, 27100 Pavia;
e-mail: camerino@unipv.it

Michael Clinton, Roslin Institute, Department of gene expression
and development, Midlothian, EH 25 9PS, UK;
e-mail: michael.clinton@bbsrc.ac.uk

Mireille Dorizzi, Institut Jacques Monod, CNRS, and Universités Paris 6
et Paris 7, 2 Place Jussieu, F-75251 Paris Cedex 05, France;
e-mail: dorizzi@ijm.jussieu.fr

Peter N. Goodfellow, Department of Genetics, University of Cambridge,
Downing Street, Cambridge CB2 3EH, UK

Yann Guiguen, INRA (Institut National de la Recherche Agronomique),
Laboratoire de Physiologie des Poissons, Campus de Beaulieu,
35042 Rennes Cedex, France; e-mail: guiguen@beaulieu.rennes.inra.fr

Lynne C. Haines, Comparative and developmental genetics section, MRC,
Human genetics unit, Western general Hospital, Edinburgh EH4 2XU, UK

Peter Koopman, Centre for Molecular and Cellular Biology,
Institute for Molecular Bioscience, The University of Queensland,
Brisbane, QLD 4072, Australia;
e-mail: p.koopman@imb.uq.edu.au

Jennifer A. Marshall Graves, School of Genetics and Evolution,
LaTrobe University, Melbourne, Vic 3083, Australia;
e-mail: J. Graves@gen.Latrobe.edu.au

Keith L. Parker, Department of Internal Medicine and Pharmacology,
UT Southwestern Medical Center, Dallas, TX 75235, USA;
e-mail: Keith.Parker@UTSouthwestern.edu

Andrew Pask, Department of Zoology, The University of Melbourne,
Parkville, Vic 3052, Australia;
e-mail: a.pask@zoology.unimelb.edu.au

Claude Pieau, Institut Jacques Monod, CNRS, and Universités Paris 6
et Paris 7, 2 Place Jussieu, F-75251 Paris Cedex 05, France;
e-mail: pieau@ijm.jussieu.fr

Noëlle Richard-Mercier, Institut Jacques Monod, CNRS, and Universités
Paris 6 et Paris 7, 2 Place Jussieu, F-75251 Paris Cedex 05, France;
e-mail: mercier@ijm.jussieu.fr

Andreas Schedl, Max-Delbrück-Centrum for Molecular Medicine,
Robert-Rössle-Strasse 10, 13125 Berlin, Germany;
e-mail: aschedl@mdc-berlin.de

Gerd Scherer, Institute of Human Genetics and Anthropology,
University of Freiburg, Breisacher Strasse 33, 79106 Germany;
e-mail: scherer@humangenetik.ukl.uni-freiburg.de

Bernard P. Schimmer, The Banting and Best Department of Medical
Research, University of Toronto, Toronto, Canada M5G 1L6;
e-mail: Bernard.Schimmer@Utoronto.ca

Michael Schmid, Department of Human Genetics, University
of Würzburg, Biozentrum, Am Hubland, 97074 Würzburg, Germany;
e-mail: m.schmid@biozentrum.uni-wuerzburg.de

Claus Steinlein, Department of Human Genetics,
University of Würzburg, Biozentrum, Am Hubland,
97074 Würzburg, Germany;
e-mail: steinlei@biozentrum.uni-wuerzburg.de

Preface

One of the most fundamental biological processes is the determination, and subsequent differentiation, of the two sexes. To achieve this "masterpiece of nature" – to borrow a phrase from Erasmus Darwin used in allusion to sexual reproduction – several mechanisms have evolved in the different vertebrate classes. Mammals, including the marsupials, have a genetic sex-determining mechanism where the male is the heterogametic sex (XY), the female the homogametic sex (XX). Birds also have a genetic sex-determining mechanism, but here it is the female that is the heterogametic sex (ZW), whereas the male is the homogametic sex (ZZ). Reptiles, amphibians and fish exhibit different mechanisms of sex determination: (1) genotypic sex determination of both the XX/XY and ZZ/ZW type whereby the sex chromosomes are either heteromorphic, as in mammals or birds, or homomorphic, identifiable only by genetic means; and (2) environmental (phenotypic) sex determination such as temperature-dependent sex determination, most intensely studied in oviparous reptiles. Another significant difference between mammals and the other vertebrate classes is that in the latter, gonadal development is under the influence of sex steroid hormones, and steroid-induced sex reversal is extensively studied in non-mammalian vertebrates.

Following the isolation of the *SRY* gene ten years ago, a handful of other genes have meanwhile been identified, mainly by positional cloning in human sex reversal syndromes, and shown to play essential roles in early gonadal development and differentiation. These include *SF1*, *WT-1*, *DAX1*, *SOX9* and, more recently, *DMRT1*. Other than *SRY*, an evolutionary newcomer found only in mammalian vertebrates, these additional genes are conserved in all vertebrates. So despite the differences in the mechanisms vertebrates use to determine sex, the same basic set of transcription factor genes appears to operate. What has also become clear is the fact that sex determination in vertebrates is not the result of a simple hierarchical cascade of gene actions as initially thought, but rather results from a complex network of positive and negative regulatory interactions. Present work largely revolves around the theme of characterizing this interactive network, with the mammalian field at the front, but with research in the other vertebrate classes gradually catching up. Although all these present studies must be somewhat limited, as only a handful of transcription factors are known, they already provide a much more comprehensive picture compared to that of just a few years ago. At the same time, the often contradictory and difficult to interpret data painfully point out our still considerable ignorance, indicating that important pieces of the puzzle are still missing.

The reviews collected in this monograph thus provide a snapshot of the status of our knowledge of the genetics and of the developmental processes in vertebrate sex determination ten years "post *SRY*". We leave it to every-one's imagination to picture the status of the field another decade hence. Certainly, more of the missing players, additional transcription factors and other molecules, will have been identified and their interconnections with the already known actors in nature's "masterplay" of sex determination will have been described. Time will tell if this future knowledge will lead to a still more unifying picture regarding gene networks common to the early steps in gonadal development in all vertebrate classes, or rather highlight the different solutions to the same problem. The stage is set. The years ahead should reveal ever more clearly the plot (or plots) behind this fas-cinating chapter of biology.

Gerd Scherer   Michael Schmid

Genes and Mechanisms in Vertebrate Sex Determination
ed. by G. Scherer and M. Schmid
© 2001 Birkhäuser Verlag Basel/Switzerland

The one-to-four rule and paralogues
of sex-determining genes

Susumu Ohno

*Beckman Research Institute of the City of Hope, 1450 East Duarte Road, Duarte,
California 91010-3000, USA*

Summary. Because of two successive rounds of tetraploidization at their inception, the vertebrates contain four times more protein-coding genes in their genome than the invertebrates:
60 000 vs. 15 000. Consequently, each invertebrate gene has been amplified to the maximum of
four paralogous genes in vertebrates: the one-to-four rule. When this rule is applied to genes
pertinent to gonadal development and differentiation, the following emerged: (i) Two closely
related zinc-finger transcription factor genes in invertebrates have been amplified to two
paralogous groups in vertebrates. One consisted of *EGR1*, *EGR2*, *EGR3* and *EGR4*, whereas
the only known paralogue of the other is *WT1*, which controls the developmental fate of the
entire nephric system, and therefore of gonads. Interestingly, *EGR1* and *WT1* act as antagonists
of each other in nephroblastic cells. (ii) *SF-1*, which controls the fate of two steroid hormone-
producing organs, adrenals and gonads, is descended from the invertebrate *Ftz-F1* gene, and its
only known paralogue is *GCNF-1*. (iii) The Y-linked *SRY*, the mammalian testis-determining
gene, is a paralogue neither of *SOX3* (*SRX*) nor of *SOX9*. Its ancient origin suggests that *SRY*
once became extinct in earlier vertebrates, only to revive itself in the mammalian ancestor.
(iv) Inasmuch as four paralogues of one invertebrate nuclear receptor gene have differentiated
to receptors of androgen, mineralocorticoid, glucocorticoid and progesterone, there should at
most be four paralogous estrogen-receptor genes in the vertebrate genome. It is likely that
one of them plays a pivotal role in the estrogen-dependent sex-determining mechanism so
commonly found among reptiles, amphibians and fish.

Introduction

The somewhat satirical Roman proverb "mutatis mutandis, ipsissima omnia"
("all the necessary changes having been made, all the things remain as
before") succinctly summarizes my long-held belief on the nature of evolution. Indeed, even after profound evolutionary changes in the body form
and organ types, the same gene more often than not performs the same function as before; thus, all the changes having been made, everything, in fact,
remained the same. The *PAX6* gene that governs the development of all
metazoan eyes can be given as the best example of the above.

Metazoan eye formation was traditionally invoked as the classical example of convergent evolution, meaning achievements of the same end by
divergent genetic means. At first glance, it indeed appears so, since animals
equipped with eyes have a peculiar way of showing up in unexpected
branches of various phyla without warning, as it were. For example, a vast
majority of the flatworms belonging to the phylum Plathelminthes are
devoid of eyes. Yet planarias of the class Turbellaria possess eyes. Among

the phylum Mollusca, members of the class Cephalopoda such as squids are uniformly equipped with eyes, whereas clams and mussels of the class Bivalvia are devoid of eyes with equal uniformity. The class Gastropoda, on the other hand, is a mixed bag containing some with and others without eyes. Furthermore, compound eyes of crustaceans and insects belonging to the phylum Arthropoda are very different from single eyes of other animals. In spite of all the above, it has finally been proven that metazoan eyes are invariably formed under the direction of the one specific gene: *PAX6* [1, 2]. In addition, the primacy of the *PAX6* gene in eye formation has been established by coupling the *PAX6* coding sequence to various regulatory elements in *Drosophila*. Eyes formed at various ectopic sites where *PAX6* was expressed [2].

Because of my belief on the nature of evolution, when I was informed by my distinguished immunologist colleague Edward A. Boyse in 1975 that S. S. Wachtel and G. C. Koo in his laboratory found H-Y plasma membrane antigen to be heterogametic sex-specific not only among vertebrates but also among invertebrates, my immediate reaction was that this H-Y antigen must be the long-sought-after universal primary determiner of the heterogametic sex [3].

As it turned out, it is the Y-linked *SRY* transcription regulator gene that directs mammalian testicular differentiation [4, 5]. Subsequently, however, it was found that, albeit a member of the *SOX* family of genes, *SRY* as such is not found in other classes of vertebrates. Furthermore, the *SRY*-dependent sex determination sensu stricto is not universal even among mammals, for males of a certain esoteric microtine rodent species, *Ellobius lutescens*, are devoid of *SRY* [6]; in fact, the entire Y chromosome is absent in this species, with males and females sharing the identical X0 sex chromosome constitution [6]. It follows that *SRY* is a violator of the magnum dictum of evolution, and as such *SRY* is irrelevant to the sex-determining mechanisms of birds, reptiles, amphibians and fish. In this paper, I would like to consider various genes involved in vertebrate sex determination in the broader context.

The one-to-four gene number rule among invertebrates and vertebrates

While a few of the ongoing genome projects on diverse species have finally been completed, others, too, have already yielded numerous relevant information. Table 1 shows that three invertebrate species representing the three different phyla Nemathelminthes, Arthropoda and Chordata are endowed with about the same total number of protein-coding genes in the genome, the number being 15000 or thereabout [7]. Particularly noteworthy is the fact that a tunicate *Ciona* is an invertebrate member of the phylum Chordata to which we vertebrates also belong. In table 1, gnathos-

Table 1. Difference in gene numbers between invertebrates and gnathostomic vertebrates

	Total number of gene loci	Genome size in numbers of base pairs
Invertebrates		
Phylum		
Nemathelminthes		
Nematode		
Caenorhabditis elegans	$17\,500 \pm 1500$	1.00×10^{8}
Phylum		
Arthropoda		
Fruit fly		
Drosophila melanogaster	$13\,500 \pm 2500$	1.65×10^{8}
Phylum		
Chordata		
Subphylum		
Urochordata		
Tunicate		
Ciona intestinalis	$15\,500 \pm 2500$	1.90×10^{8}
Gnathostomic vertebrates		
Phylum		
Chordata		
Subphylum		
Vertebrata		
Class		
Osteichthyes		
Japanese puffer		
Fugu rubripes	$86\,250 \pm 21\,500$	3.90×10^{8}
Class		
Mammalia		
Homo sapiens	$86\,250 \pm 21\,500$	3.00×10^{9}

tomic vertebrates are represented by a puffer fish and by our own species. As with all the extremely specialized teleost fish, the genome of a puffer fish, by a drastic secondary reduction, became a mere one-eighth the size of the mammalian genome. Yet, the total number of protein-coding genes contained in the genome of a puffer is about the same as that in the human, therefore, mammalian genome [8]. In Table 1, a concensus number of around 86 250 is given as the characteristic total number of protein-coding genes in all vertebrate genomes [7]. However, my estimate since 1970 has been more conservative; the realistic number for all vertebrates, excluding recent tetraploid fish and amphibians, was thought to be between 50 000 and 80 000 [9]. Inasmuch as 15 000 times 4 is 60 000, Table 1 is entirely compatible with the view expressed in 1970 that gnathostomic vertebrates underwent two successive rounds of tetraploidization at their inception [9]. In short, the vertebrate genome contains four times more gene loci than the invertebrate genome. Thanks to ever-increasing genomic information, the octaploid nature of vertebrate genomes has been receiving growing support in recent years. The most revealing was the one-to-four rule proposed by

Spring [10]. He pointed out that, because of the octaploid nature of vertebrate genomes, each single gene locus of *Drosophila*, as a representative of invertebrates, has been amplified, as a rule, to four paralogous genes in humans. This point is illustrated in Table 2, which lists eight varieties of genes with regulatory roles [10]. Needless to say, only two or three instead of all four paralogues were occasionally encountered in the human genome. For example, it would be seen in Table 2 that the gene *ci* of *Drosophila*, which encodes a zinc-finger transcription factor of the glioblastoma family, has been amplified to only three paralogues, *GLI*, *GLI2* and *GLI3*, in the human genome. Inasmuch as the human genome project is still far from completion, it is probable that a missing fourth paralogue will be found in the future. On the other hand, it will be no surprise if one or two of the original four paralogous genes have degenerated into functionless pseudogenes. After all, the last tetraploidization event is thought to have taken place 450 million years ago at the end of the Ordovician Period. The relevance of this one-to-four rule to our understanding of interactions between various known genes involved in development and differentiation of the gonad shall now be discussed.

WT1 (11p13) and its pseudo-paralogue *EGR1* (5q23–q31)

Mice homozygously-deficient for the *Wt1* (Wilms' nephroblastoma) gene die in utero due to the absence of metanephros development, and this developmental failure apparently extends to the mesonephros as well, the absence of gonads being a necessary consequence of mesonephric failure [11]. Accordingly, *WT1* resides at the top of the regulatory hierarchy governing nephric development, of which gonadal development is a part. The vertebrate genome contains a few hundred gene loci that encode numerous families of zing-finger transcription regulators. Of those, *WT1* is most closely related to the *Egr/Krox-20* family. Table 2 shows that the *sr* gene of *Drosophila* has been amplified to four *EGRs* (*early growth response* genes) in humans [10]. According to the one-to-four rule, the ready presence of *EGR1*, *EGR2*, *EGR3* and *EGR4* in the vertebrate genome implies that *WT1* is a member not of the *Egr/Krox-20* family itself but of its very close ally. Thus three paralogues of *WT1* are expected at most. Interestingly, it appears that, whereas *WT1* exerts an inhibitory effect on nephroblast proliferation, *EGR1* promotes nephroblast proliferation by antagonizing *WT1* [12]. In view of the above, the following four experiments seem worthwhile: (i) searching for true paralogues of *WT1*; (ii) knockout of *Egr1* in mice; (iii) double knockouts of *Wt1* and *Egr1* in mice; (iv) the same knockout and double knockouts in amphibians. We recall that mesonephros persists as adult kidney in amphibians.

Table 2. One-to-four rule as it applies to regulatory genes (modified from [10])

	Invertebrates	Vertebrates	
	Drosophila	human	
		gene	chromosomal location
Notch		*NOTCH1*	9q34.3
epidermal-neuronal	*N*	*NOTCH2*	1q11–p13
cell-cell interaction		*NOTCH3*	19p13
receptors		*INT3*	6p21.3
Mef2		*MEF2A*	15q25
MADS box	*Mef2*	*MEF2B*	19p12
enhancing factors		*MEF2C*	5q14
		MEF2D	1q12–q23
Ras		*RRAS*	19q13
GTP binding	*RAS85D*	*HRAS*	11p15.5
oncogenes		*KRAS2*	12p12.1
		NRAS	1p13
Egr/Krox-20		*EGR1*	5q23–q31
zinc finger	*sr*	*EGR2*	10q21.1
transcription factors		*EGR3*	8p21–p23
		EGR4	2p13
Gli		*GLI*	12q13
zinc-finger transcription factors	*ci*	*GLI2*	2
glioblastoma family		*GLI3*	7p13
Src		*SRC*	20q11.2
nonreceptor tyrosine kinase	'Src41A'	*YES1*	18p11
protooncogenes		*FGR*	1p36
		FYN	6q21
Src-related		*LCK*	1p34–p35
nonreceptor tyrosine kinases	*Src64B*	*LYN*	8q13
		HCK	20q11–q12
		BLK	8p22–p23
Jak		*JAK1*	1p31–p32
nonreceptor tyrosine kinases B	*hop*	*JAK2*	9p24
		JAK3	?
		TYK2	19p13.2

SF-1 (9q33), *GCNF-1* and their unidentified paralogues

In the homogzygous absence of the *Sf-1* gene, mice develop neither adrenals nor gonads, two steroid hormone-synthesizing organs of vertebrates [13]. Accordingly, *SF-1* occupies the second position in the regulatory hierarchy of gonadal development. It is fitting that *SF-1* also exerts transcriptional control over genes encoding steroid hormone-synthesizing enzymes. *SF-1* is one of the genes that encode orphan nuclear receptors. These nuclear receptors without ligands are extremely ancient, as shown in an extensive phylogenic study that revealed the presence of two (COUP and

Table 3. Nuclear receptor family

	Invertebrates	Vertebrates
	Drosophila	human
1. Orphan		
Coup-type		*EAR1*
	cup	*EAR2*
		EAR3 (COUP)
FTZ-F1-type		*SF-1*
	Ftz-F1	*GCNF-1*
2. Retinoic acid		
Type A		*RARA*
	E78	*RARB*
		RARG
Type X		*RXRA*
	usp	*RXRB*
		RXRG
3. Thyroid hormone, prostaglandin, vitamin D		
Thyroid hormone		*TRA*
	?	*TRB*
Prostaglandin, leukotriene		*PPARA*
	E75	*PPARG*
Vitamin D		*VDR*
	ECR (ecdysone)	*MB67*
4. Steroid hormones		
		AR (androgen)
	?	*MR (mineralocorticoid)*
		GR (glucocorticoid)
		PR (progesterone)
		ER (estrogen)
	?	*ERR1 (estrogen)*
		ERR2 (estrogen)

FTZ-F1) types of orphan nuclear receptors in diploblastic animals lacking mesoderm of the ancient and primitive phylum Cnidaria such as a sea amenone [14]. Recall that diploblastic animals belonging to the two phyla Porifera and Cnidaria were exceptional metazoans in that they appeared considerably before the Cambrian explosion that started 530 million years ago. Table 3 shows that a COUP-type orphan nuclear receptor in *Drosophila* is encoded by the *cup* gene and that this gene has been amplified to *EAR1*, *EAR2* and *EAR3* in the human genome. Of the three *EARs*, *EAR3* is a homologue of the chicken *COUP* gene. Inasmuch as the *COUP* gene in chicken was originally discovered as a transcriptional regulator of an ovalbumin gene, EARs, too, are not altogether unrelated to reproductive functions. The second type's namesake *FTZ-F1* in *Drosophila* is a transcriptional regulator of the *FTZ* gene that is involved in very early embryonic segmentation processes. One of its paralogues in vertebrates is *SF-1*, and the other is *GCNF-1*. Needless to say, the elucidation of *GCNF-1* function and the identification of two other paralogues of *SF-1* would be extremely rewarding.

The antiquity of the *SRY* (Yp11.2) gene in solitary splendor

As stated at the beginning, the *SRY* gene is found only among mammals. At first glance, this implies that *SRY* is a very recent derivative of a particular *SOX* gene paralogue. The highly schematic dendrogram of *SOX* gene families presented in Figure 1 reveals otherwise. This simplified dendrogram

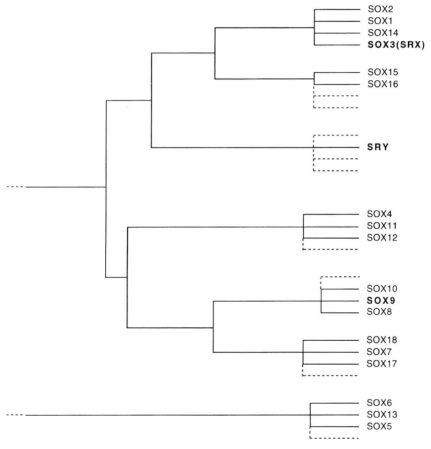

Figure 1. A highly stylized dendrogram of 19 known members of the SOX transcription factor family [15, 16]. With regard to the amino acid sequence encompassing residues 11–66 of the HMG domain, members of the same paralogous group (e.g. SOX2, SOX1, SOX14, SOX3) maintained 84% or greater identities, whereas identities between different paralogous groups were between 61% and 43%. Identities between SRY and other SOX proteins were at best 63% and at worst 47%.

was derived by combining data from two independent sources [15, 16]. Figure 1 shows that *SOX* genes of vertebrates, too, obey the one-to-four rule, which in turn reveals that there already were seven independent *SOX* gene loci in the invertebrate genome, one of which was an ancestor of *SRY*. Accordingly, *SRY* is closely related neither to *SOX3* nor to *SOX9*. Recall that *SOX3* is also known as *SRX*, as its human form resides on Xq26–q27. Clearly, *SOX3* and *SRY* have never been alleles of each other. Recall also that human individuals heterozygous for a defective *SOX9* gene (17q24–q25) manifest a form of osteochondro-dysplasia known as campomelic dysplasia accompanied by XY sex reversal [17]. Figure 1 shows that para-logues of *SOX9* are *SOX8* and *SOX10* but not SRY. In view of the antiquity of the *SRY* gene, an extensive search for its ancestor among various inver-tebrates emerges as an extremely worthwhile undertaking. If *SRY* is truly absent in vertebrate classes other than Mammalia, the *SRY* gene must be one of those genes that became extinct once and later revived, in this par-ticular case in the mammalian ancestor. The trichrome color vision that depends upon the presence of three opsins (red, green and blue) in cone photoreceptor cells of the retina is well developed in fish as well as in birds, but color vision was secondarily lost in mammals, probably because the first mammal to emerge under the shadow of the dinosaurs was a minute nocturnal insectivore of the Cretaceous Period some 100 million years ago [18]. This loss was caused by defective mutations sustained at the two X-linked gene loci; a red opsin gene began to encode a degenerate protein, whereas a green opsin gene was extinguished. Restoration of trichrome color vision in higher primates including humans was due to rebuilding X-linked red and green opsin genes from a once degenerate red opsin gene. Indeed, a new red opsin gene and a new green opsin gene initially re-appeared as alleles of a single locus [19].

The very fact that even a mammalian species can operate the sex-deter-mining mechanism of the male heterogamety without the benefit of *SRY* [6] suggests that there is a gene which is an antagonist of *SRY*. As already noted, whereas *WT1* functions as a transcriptional repressor in nephroblastic cells, EGR1 acts as a transcription activator [12]. In the above type of antago-nism, either the heterozygous deficiency of an autosomal antagonist gene or the hemizygous absence of an X-linked antagonist gene might recreate the condition normally brought about by the presence of *SRY*.

The role of sex steroid nuclear receptors in hormone-dependent sex-determining mechanisms

As Claude Pieau and colleagues point out in great detail in their own review, temperature-dependent sex-determining mechanisms widely prac-ticed by alligators, turtles and lizards of the class Reptilia are apparent con-sequences of varying degrees of temperature sensitivities exhibited by individual enzymes involved in synthesis of sex steroids. At the simplest, a species whose steroid aromatase does not function well at a lower incubation temperature would produce males at that temperature simply becuse not enough androstenedione and testosterone are converted to oestrone and oestradiol. In fact, gonadal development of all vertebrates, excepting mammals, is under the influence of sex steroid hormones. It follows that in the vast majority of vertebrates, the pivotal role in sex determination is played by a particular branch of the above-discussed family of nuclear receptor transcription factors that utilize sex steroids as ligands. Evolutionarily as ancient as orphan nuclear receptors are retinoic acid nuclear receptors, since they were already found together with the former in diploblastic animals of the phylum Cnidaria [14]. As shown in Table 3, the *Drosophila* genome as a representative of invertebrate genomes contains one ancestral gene each encoding type A and type X retinoic acid nuclear receptors. In vertebrates, both have been amplified, the former to *RARA*, *RARB* and *RARG*, and the latter to *RXRA*, *RXRB* and *RXRG* [10, 14]. Next in the evolutionary order comes a mixed bag of least-explored nuclear receptors that utilize thyroid oligopeptidic hormone, vitamin D and prostaglandins as well as leukotrienes as ligands. Ancestral genes for some but not all of these nuclear receptors in this mixed bag have been identified in the invertebrate genome. For example, *Drosophila E75* is an apparent ancestor of *PPARA* and *PPARG*, and ecdysone (insect metamorphosis hormone) receptor gene *Ecr* of *Drosophila* is ancestral to vitamin D receptor paralogues *VDR* and *MB67*, as shown in Table 3 [14]. Nuclear receptors that utilize steroid hormones as ligands are present only in vertebrates, and Table 3 shows that they form two independent paralogous groups. Androgen receptor *AR*, mineralocorticoid receptor *MR*, glucocorticoid receptor *GR* and progesterone receptor *PR* are four paralogues descended from one ancestral invertebrate gene. The other ancestral gene has been amplified to three known paralogous estrogen receptors in vertebrates. Such a redundancy has likely created functional differentiation among *ER* paralogues. In certain vertebrates, but not in mammals, one such *ER* paralogue might have begun to control the transcription of a set of genes responsible for differentiation toward an ovary of the indifferent gonad thereby creating the estrogen-dependent autocrine sex-determining mechanism.

References

1 Matsuo T, Osumi-Yamashita N, Noji S, Ohuchi H, Koyama E, Myokai F et al (1993) A mutation in the Pax-6 gene in rat *small* eye is associated with impaired migration of mid-brain crest cells. *Nature Genet* 3: 299–304
2 Halder G, Callaerts P and Gehring WJ (1995) Induction of ectopic eyes by targeted expression of the *eyeless* gene in *Drosophila*. *Science* 267: 1788–1792
3 Wachtel SS, Ohno S, Koo GC and Boyse EA (1975) Possible role of H-Y antigen in primary determination of sex. *Nature* 257: 235–236

4 Sinclair AH, Berta P, Palmer MS, Hawkins JR, Griffiths BL, Smith MJ et al (1990) A gene from the human sex-determining region encodes a protein with homology to a conserved DNA-binding motif. *Nature* 346: 240–244

5 Koopman P, Gubbay J, Vivian N, Goodfellow P and Lovell-Badge R (1991) Male development of chromosomally female mice transgenic for *Sry*. *Nature* 351: 117–121

6 Just W, Rau W, Vogel W, Akhverdian M, Fredga K, Graves JAM et al (1995) Absence of *Sry* in species of the vole *Ellobius*. *Nature Genet* 11: 117–118

7 Simmen MW, Leitgeb S, Clark VH, Jones SJM and Bird A (1998) Gene number in an invertebrate chordate, *Ciona intestinalis*. *Proc Natl Acad Sci USA* 95: 4437–4440

8 Brenner S, Elgar G, Sandford R, Macrae A, Venkatesh B and Aparicio S (1993) Characterization of the pufferfish (*Fugu*) genome as a compact model vertebrate genome. *Nature* 366: 265–268

9 Ohno S (1970) Evolution by Gene Duplication, Springer, Berlin

10 Spring J (1997) Vertebrate evolution by interspecific hybridisation – are we polyploid? *FEBS Lett* 400: 2–8

11 Kreidberg JA, Sariola H, Loring JM, Maeda M, Pelletier J, Housman D et al (1993) WT-1 is required for early kidney development. *Cell* 74: 679–691

12 Madden SL, Cook DM, Morris JF, Gashler A, Sukhatme VP and Rauscher III FJ (1991) Transcriptional repression mediated by the WT1 Wilms tumor gene product. *Science* 253: 1550–1553

13 Luo X, Ikeda Y and Parker KL (1994) A cell-specific nuclear receptor is essential for adrenal and gonadal development and sexual differentiation. *Cell* 77: 481–490

14 Escriva H, Safi R, Haenni C, Langlois M-C, Saumitou-Laprade P, Stehelin D et al (1997) Ligand binding was acquired during evolution of nuclear receptors. *Proc Natl Acad Sci USA* 94: 6803–6808

15 Wright EM, Snopek B and Koopman P (1993) Seven new members of the *Sox* gene family expressed during mouse development. *Nucleic Acids Res* 21: 744

16 Stock DW, Buchanan AV, Zhao Z and Weiss KM (1996) Numerous members of the Sox family of HMG box-containing genes are expressed in developing mouse teeth. *Genomics* 37: 234–237

17 Meyer J, Suedbeck P, Held M, Wagner T, Schmitz ML, Bricarelli FD et al (1997) Mutational analysis of the *SOX9* gene in campomelic dysplasia and autosomal sex reversal: lack of genotype/phenotype correlations. *Hum Mol Genet* 6: 91–98

18 Ohno S (1967) Sex Chromosomes and Sex-Linked Genes, Springer, Berlin

19 Jacobs GH and Neitz J (1987) Inheritance of color vision in a New World monkey (*Saimiri sciureus*). *Proc Natl Acad Sci USA* 84: 2545–2549

Genes and Mechanisms in Vertebrate Sex Determination
ed. by G. Scherer and M. Schmid
© 2001 Birkhäuser Verlag Basel/Switzerland

Genes essential for early events in gonadal development

Keith L. Parker[1], Bernard P. Schimmer[2] and Andreas Schedl[3]

[1] *Departments of Internal Medicine and Pharmacology, UT Southwestern Medical Center, Dallas, TX 75235, USA*
[2] *The Banting and Best Department of Medical Research, University of Toronto, Toronto, Ont. M5G 1L6 Canada*
[3] *Max-Delbrück-Centrum for Molecular Medicine, Robert-Rössle-Str. 10, D-13125 Berlin, Germany*

Summary. The acquisition of a sexually dimorphic phenotype is a critical event in mammalian development. The basic underlying principle of sexual development is that genetic sex – determined at fertilization by the presence or absence of the Y chromosome – directs the embryonic gonads to differentiate into either testes or ovaries. Thereafter, hormones produced by the testes direct the developmental program that leads to male sexual differentiation. In the absence of testicular hormones, the female pathway of sexual differentiation occurs. Recent studies have defined key roles in gonadal development for two transcription factors: Wilms' tumor suppressor 1 (WT1) and steroidogenic factor 1 (SF-1). After presenting a brief overview of gonadal development and sexual differentiation, this chapter reviews the studies that led to the isolation and characterization of WT1 and SF-1, and then discusses how interactions between these two genes may mediate their key roles in a common developmental pathway.

Introduction

Prior to sexual differentiation, the ovaries and testes cannot be distinguished and therefore are called bipotential or indifferent gonads. These bipotential gonads arise from the urogenital ridge, a region adjacent to the mesonephros that ultimately contributes cell lineages to the adrenal cortex, gonads and kidney. The testes and ovaries have functional counterparts that serve corresponding functions in reproduction. These counterparts include the Leydig and theca cells, which comprise the steroidogenic compartment, the Sertoli and granulosa cells, which support germ cell maturation, the germ cells (spermatocytes and oocytes), and the peritubular myoid and stroma cells – which form the connective tissue of the gonads.

After sexual determination, the testes and ovaries can be distinguished histologically, largely because the testes organize into two distinct compartments: the testicular cords and the interstitial region. The testicular cords – precursors of the seminiferous tubules – contain the fetal Sertoli cells and the primordial germ cells, which migrate into the gonad from a position outside of the urogenital ridge. The interstitial region, surrounding the testicular cords, contains the steroidogenic Leydig cells and the peritubular myoid cells. In contrast, the ovaries have an amorphous, "ground-glass" appearance and exhibit little structural differentiation until late in gestation.

The internal genitalia derive from the genitourinary tract, which again initially is identical in male and female embryos. At the indifferent stage, male and female embryos have two identical sets of paired ducts: the Müllerian (paramesonephric) ducts and the Wolffian (mesonephric) ducts. If the Y chromosome activates the male developmental pathway, testes develop and ultimately effect male sexual differentiation by causing the Müllerian ducts to degenerate and the Wolffian ducts to develop into the seminal vesicles, epididymis, and vas deferens. In the absence of testicular hormones, the Wolffian ducts regress and the Müllerian ducts form the oviducts, Fallopian tubes, uterus, and upper vagina. As predicted by the classic studies of Jost [1, 2], the critical mediator of Müllerian duct regression in males is a glycoprotein hormone, Müllerian inhibiting substance (MIS), which is produced by Sertoli cells within the testicular cords. Testicular androgens, synthesized by Leydig cells in the interstitial region, cause male differentiation of the Wolffian ducts and external genitalia.

The external genitalia, like the internal genitalia, also derive from structures that initially are found in both sexes, including the genital tubercle, urethral folds, the urethral groove, and the genital swellings. Again, androgens are critical for virilization, although full virilization of the external genitalia requires the conversion of testosterone to dihydrotestosterone by 5α-reductase [3].

As summarized in the chapter by Koopman, we now know that a single gene, designated *SRY* for *sex-determining region-Y chromosome*, mediates male sexual determination (reviewed in [4]). However, the mechanisms by which *SRY* activates the male pathway remain undefined. The structural homology of SRY to transcriptional regulators of the high-mobility-group family led to the hypothesis that SRY activates downstream genes, which in turn mediate the conversion of the bipotential gonad into a testis. Direct target genes of SRY, however, have not been isolated, and the role of SRY as a regulator of gene transcription thus remains to be proven. In an effort to extend our understanding of sexual differentiation, other groups have tried to identify additional genes that play important roles in early stages of gonadal development and sexual differentiation.

Genes essential for early stages of gonadal development

From the above discussion, it is apparent that mutations in genes that are essential for development of the bipotential gonad – before sexual determination and differentiation take place – will impair testes formation in an XY background or ovary formation in an XX background. Recent studies have shown that two different transcription factors, WT1 and SF-1, play such pivotal roles in early gonadogenesis.

WT1: a critical mediator of urogenital development

The *WT1* gene initially was isolated through analyses of patients with Wilms' tumors, an embryonic kidney tumor arising from abnormal proliferation of the metanephric blastema (reviewed in [5]). Although Wilms' tumor generally presents sporadically, approximately 1% of Wilms' tumors occur in patients with similarly affected first-degree relatives. This finding led to the proposal that a tumor suppressor gene was mutated in these families in a manner akin to retinoblastoma. Efforts to map the gene(s) were largely guided by heterozygous deletions on human chromosome 11p13, which are associated with the WAGR syndrome [6]. Patients with this syndrome exhibit a variable phenotype that includes Wilms' tumors, aniridia, genitourinary abnormalities, and mental retardation. Genitourinary abnormalities are observed only in a subset of patients and are relatively mild, involving cryptorchidism and hypospadias in males and horseshoe kidneys in males and females. The WAGR phenotype reflects deletions of several genes, including *WT1* [7–9] and the transcription factor *PAX6* [10], isolated mutations of which also are associated with aniridia in humans. Two other genes within the chromosomal region deleted in WAGR patients are expressed in the embryonic brain and therefore may contribute to the mental retardation phenotype [11, 12]. A small percentage of patients with familial Wilms' tumors inherit mutations or deletions of one WT1 allele, and then undergo somatic loss of the second allele due to gross chromosomal events. Other chromosomal regions associated with familial Wilms' tumors include 11p15 (the region associated with the Beckwith-Wiedemann syndrome) and chromosome 17. These other Wilms' tumor genes have not yet been identified, and remain an ongoing area of investigation.

WT1, by alternative splicing [13] and alternative translation start sites [14, 15], generates at least twelve different isoforms of a zinc-finger DNA-binding protein that is thought to regulate gene transcription by interacting with specific DNA recognition sequences upstream of target genes. Besides differing somewhat in their preferences for DNA binding [16], the different isoforms of WT1 also localize differentially within the nucleus. *In vitro* experiments demonstrated that the localization is controlled mainly by the presence or absence of the second alternative splice, which introduces three amino acids (lysine-threonine-serine, KTS) between zinc fingers 3 and 4. Whereas –KTS isoforms show a more diffuse nuclear staining, +KTS isoforms associate with spliceosomes, suggesting that they may participate in RNA processing [17–19]. Studies to date have not yielded any evidence for tissue-related differences in the relative levels of the different WT1 isoforms and the molecular function of these WT1 isoforms remains unclear.

The first indication of an essential role for *WT1* in urogenital development came from analyses of its expression, which showed specific staining within the developing kidneys and gonads [20]. Subsequent analyses showed that point mutations in the *WT1* gene can lead to mild abnormalities of the genital system including hypospadias and cryptorchidism [21]. A clear role for *WT1* in development of the kidneys and gonads was established by analyses of *WT1* knockout mice [22]. In addition to renal agenesis, these *WT1* knockout mice lacked gonads, and had impaired adrenal and spleen development [23, 24]. As a result of their gonadal dysgenesis before the time that androgens and MIS are produced, the internal and external genitalia developed along the female program ([22] and our own unpublished results). These results, coupled with the structural similarity of WT1 with other transcription factors, suggest that WT1 regulates the expression of target genes that are essential for gonadogenesis in both males and females.

Besides the classical Wilms' tumor and the WAGR syndromes, *WT1* mutations are associated with two other clinical syndromes in human patients – Denys-Drash syndrome and Frasier syndrome. Denys-Drash syndrome is an autosomal dominant disorder characterized by gonadal and urogenital abnormalities in conjunction with diffuse mesangial sclerosis. The renal disease in Denys-Drash patients is quite severe, usually presenting in the first year of life and causing end-stage renal disease by age 3. The gonadal abnormalities of these patients vary, but generally are more severe than those associated with the WAGR syndrome, with streak gonads and sex-reversal of external and internal genitalia at one extreme and varying degrees of pseudohermaphroditism in less-severely affected XY males. Wilms' tumors are commonly seen in patients with Denys-Drash mutations. Denys-Drash syndrome typically results from point mutations in the zinc-finger region that abrogate DNA binding; these mutated proteins are predicted to act in a dominant negative fashion to inhibit function of the protein encoded by the wild-type allele [25–27]. In agreement with this notion it has been shown that WT1 is able to form homodimers at least *in vitro* [28, 29]. Extrapolating from the knockout mouse studies described above, it is probable that the degree of inhibition of *WT1* action correlates with the impairment of genitourinary development, with the most severe mutations leading to early gonadal dysgenesis and sex reversal of external and internal genitalia. It should, however, be noted that genetic modifiers also play important roles in the severity of the resulting phenotype. This is demonstrated in case studies of DDS patients where fathers were phenotypically normal despite carrying the same *WT1* mutation seen in affected offspring. The presence of genetic modifiers for *Wt1* has also been demonstrated in *Wt1* knock-out mice, which on a different genetic background, survive until birth without the originally observed heart defects [24].

WT1 mutations also have been identified in patients with Frasier syndrome [30–32]. Unlike patients with Denys-Drash syndrome, these patients do not develop Wilms' tumors, but instead present with gonadal dysgenesis, male pseudohermaphroditism, and focal glomerular sclerosis. Their glomerulopathy is less severe than that associated with Denys-Drash

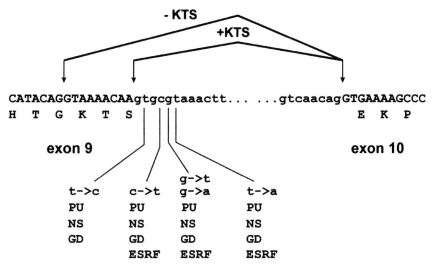

Figure 1. Mutations associated with the Frasiersyndrome disrupt splicing at the second alternative splice donor site at the end of exon 9. PU, proteinuria; NS, nephrotic syndrome; GD, gonadal dysgenesis; ESRF, end-stage renal failure.

mutations, with no evidence of renal insufficiency until after age 4 and preservation of some renal function until adolescence or young adulthood. The *WT1* mutations that cause Frasier syndrome cluster within intron 9 of the *WT1* gene, and apparently interfere selectively with the synthesis of splice variants of WT1 that insert the amino acids lysine-threonine-serine (+ KTS) between the third and fourth zinc fingers (Fig. 1). Interestingly, Frasier syndrome is a dominant disease and the wild-type allele still produces both – KTS and + KTS isoforms. These findings imply that the ratio of + KTS and – KTS isoforms is very important and that subtle changes in this ratio can cause severe developmental abnormalities. Although the significance of the WT1 splice variants is not fully understood, this finding suggests that the + KTS isoform is essential for gonadal development, but is not required to suppress the development of Wilms' tumors. Furthermore, the same *WT1* mutations that cause Frasier syndrome in boys may also cause some cases of focal glomerular sclerosis in girls, who escape diagnosis because they lack abnormalities of the external genitalia [31, 33, 34].

Steroidogenic factor 1 (SF-1): a critical mediator of endocrine development

SF-1 was described initially as an important regulator of the tissue-specific expression of the cytochrome P-450 steroid hydroxylases – enzymes that catalyze many key reactions in steroidogenesis [35, 36]. The subsequent isolation of a complementary DNA (cDNA) encoding *SF-1* showed that this critical regulator of the steroidogenic enzymes was itself a member of the nuclear hormone receptor family – proteins that mediate transcriptional activation by steroid hormones, thyroid hormone, vitamin D, and retinoids [37]. Since its initial characterization, a number of groups have shown that SF-1 regulates adrenal and gonadal expression of genes required for steroidogenesis, including the steroid hydroxylases, 3β-hydroxysteroid dehydrogenase, the adrenocorticotropin receptor and the steroidogenic acute regulatory protein (reviewed in [38]). Analyses in transfected Sertoli cells and transgenic mice further suggest that SF-1 regulates the *MIS* gene [39–41]. In addition to the steroidogenic organs, *SF-1* transcripts also were detected in the anterior pituitary and hypothalamus. Collectively, these findings suggest that SF-1 regulates the expression of both hormones that are critical for male sexual differentiation (androgens and MIS) and also raise the possibility that SF-1 plays additional roles at other levels of the endocrine axis.

Analyses of *Sf-1* knockout mice dramatically confirmed essential roles of SF-1 at all three levels of the hypothalamic-pituitary-steroidogenic organ axis. Perhaps most strikingly, these Sf-1 knockout mice lacked adrenal glands and gonads, undergoing loss of the primordial organs via programmed cell death at discrete stages of development when sexual differentiation normally takes place [42, 43]. These findings, which resemble closely the consequences of *Wt1* knockout on gonadal development, demonstrate unequivocally that Sf-1 has essential roles in the early development of the adrenal and gonadal precursors. Consistent with the degeneration of testes before androgens and MIS are produced, *Sf-1* knockout mice also exhibit male-to-female sex reversal of the internal and external genitalia. They also have impaired expression of a number of markers of pituitary gonadotropes [44, 45], the pituitary cell type that regulates gonadal steroidogenesis. Finally, they lack the ventromedial hypothalamic nucleus [45, 46], a cell group in the medial hypothalamus linked to ingestive and reproductive behaviors [47].

The early embryonic expression of *SF-1* in the gonads and its established role as a transcription factor make it likely that *SF-1* is part of the hierarchical regulatory pathway that determines the expression of downstream genes required for gonadogenesis. The human gene encoding SF-1 shares extensive homology with its mouse counterpart [48, 49] and is expressed in many of the same sites [50]. It is therefore not surprising that a *SF-1* mutation has been found in a patient with adrenocortical insufficiency and XY sex reversal [51].

Genetic interactions in gonadal development

Sexual determination and differentiation require a complex set of events in the appropriate tissues at appropriate times of development. Defects in any

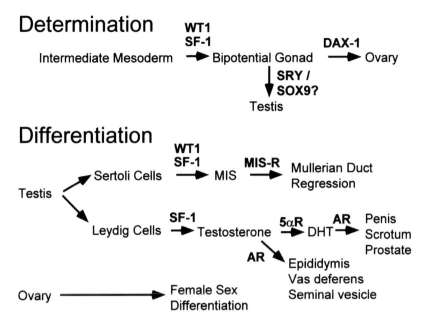

Figure 2. Summary of the molecular events in mammalian sex determination and differentiation. The positions of a number of genes believed to mediate key events in sex determination and differentiation are indicated, as discussed in the text. MIS, Müllerian inhibiting substance; MIS-R, Müllerian inhibiting substance receptor; 5αR, 5α-reductase; DHT, dihydrotestosterone; AR, androgen receptor.

of these steps can impair sexual differentiation. Although we know that SRY, encoded by the Y chromosome, is the primary mediator of male sex determination, considerable gaps remain in our understanding of just how SRY brings about these critical events in development. As summarized in Figure 2, both X-linked genes (e.g. *DAX-1*) and autosomal genes (e.g. *SF-1* and *WT1*) also play critical roles in processes of sex determination and differentiation. An important goal for future studies is to define how these genes interact in a common developmental pathway to bring about these critical developmental events.

Ongoing studies are examining potential interactions among these genes that may explain how they cooperate in gonadogenesis. Intriguingly, recent evidence supports direct functional interactions of *WT1* and *SF-1*, as well as interactions of *SF-1* with other genes implicated in gonadal and adrenal development. When analyzed *in vitro* with recombinantly expressed proteins or in mammalian two-hybrid assays, WT1 and SF-1 can form heterodimers [52]. The functional significance of this interaction is supported by the finding that cotransfection with WT1 markedly augments SF-1-dependent transcriptional activation of the *MIS* promoter. This effect was most pronounced with the – KTS isoform of WT1, a puzzling result in light of the apparent link between the +KTS form and gonadogenesis suggested by patients with Frasier syndrome. Alternatively, others have shown that cotransfection with WT1 increases the expression of reporter genes driven by the *SF-1* promoter, suggesting that WT1 may act in part to increase the levels of SF-1 (D. Lala and K. Parker, unpublished observation). In these studies, the +KTS isoform induced promoter activity most potently, providing a possible link with the abnormal testes development in patients with Frasier syndrome. It is apparent that *WT1* is not absolutely essential for *SF-1* expression, as *Wt1* knockout mice still have detectable *Sf-1* transcripts in the degenerating gonads (K. Parker and J. Kreidberg, unpublished observation). Moreover, these two models are not necessarily antagonistic, and it remains plausible that *WT1* acts both to increase levels of *SF-1* transcripts and to facilitate its activation of downstream genes such as *MIS*. To further complicate the activation of the *MIS* gene, it has been demonstrated that, in addition to WT1, the SOX9 protein is also able to stimulate the *MIS* promoter in the presence of SF-1 *in vitro* [53]. The importance of the SOX9 binding site upstream of the *MIS* promoter has recently been confirmed *in vivo* using gene targeting experiments in mouse ES-cells. Whereas mutations affecting the SF-1 binding site reduced MIS expression, mutations within the SOX9 binding site abolished MIS expression entirely [54]. Therefore, SF-1 together with WT1 may be necessary to boost and possibly stabilize MIS expression once it has been activated by SOX9. Another gene that interacts with SF-1 in endocrine development is *DAX-1* (see chapter by Goodfellow and Camerino). *DAX-1* encodes an atypical member of the nuclear receptor family that retains the conserved ligand binding domain but lacks the typical zinc-finger DNA binding motif [55], suggesting that DAX-1 regulates gene expression through protein-protein interactions. *DAX-1* was isolated initially by positional cloning of the gene responsible for X-linked adrenal hypoplasia congenita (AHC), a disorder in which patients present with ACTH-insensitive adrenal insufficiency due to impaired development of the adrenal cortex. If kept alive with corticosteroids, these AHC patients later may exhibit features of hypogonadotrophic hypogonadism, reflecting a mixed phenotype of hypothalamic and pituitary gonadotropin deficiencies. The association of impaired adrenal development and hypogonadotrophic hypogonadism resembles somewhat the phenotype in *Sf-1* knockout mice, suggesting that *DAX-1* and *SF-1* also may act in the same developmental pathway.

In support of this model, recent studies have shown that both genes are expressed in many of the same sites during embryogenesis, including the gonads, adrenal cortex, pituitary gonadotropes, and the ventromedial hypothalamic nucleus (VMH) [56, 57]. Moreover, recent studies suggest several mechanisms by which SF-1 and DAX-1 may interact (Fig. 3). One model (Fig. 3B) proposes that DAX-1 can heterodimerize with SF-1, and that DAX-1 inhibits SF-1-mediated transcriptional activation because of this heterodimerization [58]. Related studies (Fig. 3C) suggest that DAX-1 inhibits the expression of SF-1-dependent target genes by recruiting the

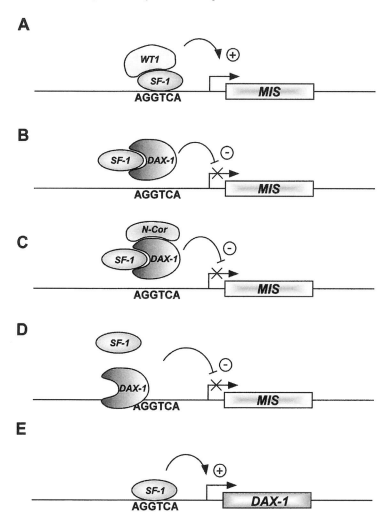

Figure 3. Potential interactions between SF-1 and DAX-1 in endocrine development. Models by which SF-1 and DAX-1 may interact in endocrine development are shown. Note that an additional SOX9 binding site mapping upstream of the SF-1 promoter element (not shown in this figure) has been demonstrated to be essential for activation of the *MIS* promoter (*A*). SF-1 induces transcription of the MIS gene by interactions with an AGGTCA promoter element. WT1 is likely to act as a stimulating cofactor in this activation process. (*B*) SF-1-DAX-1 heterodimerization inhibits activation of MIS. (*C*) DAX-1 recruits the corepressor N-Cor to SF-1-responsive promoters. (*D*) DAX-1 interacts with hairpin loops in SF-1-responsive promoters to interfere with SF-1-dependent transcriptional activation. (*E*) SF-1 activates the DAX-1 promoter.

corepressor N-Cor to their promoters [59]. Alternatively, it has been proposed (Fig. 3D) that DAX-1 interferes with SF-1 action by binding to hairpin loops in the 5′-flanking region of SF-1-responsive genes, presumably blocking access of the promoters to SF-1 [60]. Finally, there are reports that SF-1 can interact with promoter elements upstream of *DAX-1* to induce its expression (Fig. 3E), thereby providing a cooperative link between these two genes [61, 62].

The functional consequences of these proposed interactions between DAX-1 and SF-1 appear to differ depending on the tissue. In the adrenal cortex, gonadotropes, and the VMH, SF-1 and DAX-1 may cooperate to activate the expression of target genes required for tissue-specific functions. Consistent with this, the phenotypes of *Sf-1* knockout mice and *DAX-1* patients in these sites are generally concordant. In contrast, the actions of SF-1 and DAX-1 in gonadal cells appear to be antagonistic. SF-1 is required for testes development and male sexual differentiation, whereas its expression in the ovaries diminishes coincident with sexual differentiation [39, 40]. These findings suggest that SF-1 is essential for normal male sexual differentiation, but may impair ovarian development and female sexual differentiation. These findings lead to the proposal that a presumptive excess of DAX-1 – in patients with dosage-sensitive sex reversal – would suppress SF-1 function and favor ovarian development, whereas the complete absence of DAX-1 would not impede SF-1 action and therefore would be compatible with normal testicular differentiation. Interestingly, *DAX-1* has recently been suggested to be a target gene of WT1 [63]. – KTS, but not + KTS, isoforms acted as potent transcriptional activators in *in vitro* assays. These observations are also an attractive explanation for the male to female sex reversal in Frasier syndrome, in which the – KTS variants are the predominant isoforms of WT1. This excess of – KTS forms may result in an abnormal stimulation of *DAX-1* transcription and a concomitant repression of the *MIS* gene leading to the development of female genitalia.

Summary

From the studies reviewed here, it is apparent that a number of the critical genes that mediate gonadal development and sexual differentiation have now been identified. Through a combination of studies in experimental model systems (e.g. knockout mouse and transgenic overexpression studies) and analyses of additional human patients with aberrant sex determination and/or differentiation, an improved understanding of these essential developmental pathways hopefully soon will emerge.

References

1 Jost A (1953) Studies on sex differentiation in mammals. *Recent Prog Horm Res* 8: 379–418

2 Jost A, Vigier B, Prepin J and Perchellet J (1973) Studies on sex differentiation in mammals. *Recent Prog Horm Res* 29: 1–41

3 Wilson JD, Griffin JE, Russell DW (1993) Steroid 5 alpha-reductase 2 deficiency. *Endocr Rev* 14: 577–593

4 Russell DW, Goodfellow PN, Lovell-Badge R (1993) SRY and sex determination in mammals. *Annu Rev Genet* 27: 71–92

5 Hastie ND (1993) The genetics of Wilms' tumor – a case of disrupted development. *Annu Rev Genet* 28: 523–558

6 Miller RW, Fraumeni JF and Manning MD (1964) Association of Wilms' tumor with aniridia, hemihypertrophy and other congenital anomalies. *N Engl J Med* 270: 922–927

7 Call K, Glaser T, Ito C, Buckler AJ, Pelletier J, Haber DA et al (1990) Isolation and characterization of a zinc finger polypeptide gene at the human chromosome 11 Wilms' tumor locus. *Cell* 60: 509–520

8 Gessler M, Poustka A, Cavenee W, Neve RL, Orkin SH and Bruns GA (1990) Homozygous deletion in Wilms' tumours of a gene identified by chromosome jumping. *Nature* 343: 774–778

9 Haber DA, Buckler AJ, Glaser T, Call KM, Pelletier J, Sohn RL et al (1990) An internal deletion within an 11p13 zinc finger gene contributes to the development of Wilms' tumor. *Cell* 61: 1257–1269

10 Ton CC, Hirvonen H, Miwa H, Weil MM, Monaghan P, Jordan T, et al (1991) Positional cloning and characterization of a paired box- and homeobox-containing gene from the aniridia region. *Cell* 67: 1059–1074

11 Schwartz F, Eisenman R, Knoll J, Gessler M and Bruns G (1995) cDNA sequence, genomic organization, and evolutionary conservation of a novel gene from the WAGR region. *Genomics* 29: 526–532

12 Kent J, Lee M, Schedl A, Boyle S, Fantes J, Powell M et al (1997) The reticulocalbindin gene maps to the WAGR region in man and to the Small eye Harwell deletion in mouse. *Genomics* 42: 260–267

13 Haber DA, Sohn RL, Buckler AJ, Pelletier J, Call KM and Housman DE (1991) Alternative splicing and genomic structure of the Wilms tumor gene WT1. *Proc Natl Acad Sci USA* 88: 9618–9622

14 Bruening W and Pelletier J (1996) A non-AUG translation initiation event generates novel WT1 isoforms. *J Biol Chem* 271: 8646–8654

15 Scharnhorst V, Dekker P, van der Eb AJ and Jochemsen AG (1999) Internal translation initiation generates novel WT1 protein isoforms with distinct biological properties. *J Biol Chem* 274: 23456–23462

16 Bickmore WA, Oghene K, Little MH, Seawright A, van Heyningen V and Hastie ND (1992) Modulation of DNA binding specificity by alternative splicing of the Wilms' tumour wt1 gene transcript. *Science* 257: 235–237

17 Larsson SH, Charlieu JP, Miyagawa K, Engelkamp D, Rassoulzadegan M, Ross A et al (1995) Subnuclear localization of WT1 in splicing or transcription factor domains is regulated by alternative splicing. *Cell* 81: 391–401

18 Englert C, Vidal M, Maheswaran S, Ge Y, Ezzell R, Isselbacher KJ et al (1995) Truncated WT1 mutants alter the subnuclear localization of the wild-type protein. *Proc Natl Acad Sci USA* 92: 11960–11964

19 Davies RC, Calvio C, Bratt E, Larsson SH, Lamond AI and Hastie ND (1998) Wt1 interacts with the splicing factor U2AF65 in an isoform-dependent manner and can be incorporated into spliceosomes. *Genes & Dev* 12: 3217–3225

20 Pritchard-Jones K, Fleming S, Davidson D, Bickmore W, Porteous D, Gosden C et al (1990) The candidate Wilms' tumor gene is involved in genitourinary development. *Nature* 346: 194–197

21 Pelletier J, Bruening W, Li FP, Haber DA, Glaser T and Housman DE (1991) WT1 mutations contribute to abnormal genital system development and hereditary Wilms' tumor. *Nature* 353: 431–434

22 Kreidberg JA, Sariola H, Loring JM, Maeda M, Pelletier J, Housman D et al (1993) WT-1 is required for early kidney development. *Cell* 74: 679–691

23 Moore A, McInnes L, Kreidberg J, Hastie ND and Schedl A (1999) YAC complementation shows a requirement for Wt1 in the development of epicardium, adrenal gland and throughout nephrogenesis. *Development* 126: 1845–1857

24 Herzer U, Crocoll A, Barton D, Howells N and Englert C (1999) The Wilms tumor suppressor gene wt1 is required for development of the spleen. *Curr Biol* 9: 837–840

25 Pelletier J, Bruening W, Kashtan CE, Mauer SM, Manivel JC, Striegel JE et al (1991) Germline mutations in the Wilms' tumor suppressor gene are associated with abnormal urogenital development in Denys-Drash syndrome. *Cell* 67: 437–447

26 Bruening W, Bardeesy N, Silverman BL, Cohn RA, Machin GA, Aronson AJ et al (1992) Germline intronic and exonic mutations in the Wilms' tumour gene (WT1) affecting urogenital development. *Nature Genet* 1: 144–148

27 Hastie ND (1993) Dominant negative mutations in the Wilms' tumour (WT1) gene cause Denys-Drash syndrome – proof that a tumour-suppressor gene plays a crucial role in normal genitourinary development. *Hum Mol Genet* 1: 293–295

28 Moffet P, Bruening W, Nakgama H, Bardeesy N, Housman D, Housman DE et al (1995) Antagonism of WT1 activity by protein self-association. *Proc Natl Acad Sci USA* 92: 11 105–11 109

29 Reddy JC, Morris JC, Wang J, English MA, Haber DA, Shi Y et al (1995) WT-1-mediated transcriptional activation is inhibited by dominant negative mutant proteins. *J Biol Chem* 270: 10 878–10 884

30 Barbaux S, Niaudet P, Gubler MC, Grunfeld JP, Jaubert F, Kuttenn F et al (1997) Donor splice-site mutations in WT1 are responsible for Frasier syndrome. *Nature Genet* 17: 467–470

31 Klamt B, Koziell A, Poulat F, Wieacker P, Scambler P, Berta P et al (1998) Frasier syndrome is caused by defective alternative splicing of WT1 leading to an altered ratio of WT1 +/– KTS splice isoforms. *Hum Mol Genet* 7: 709–714

32 Kikuchi H, Takata A, Akasaka Y, Fukuzawa R, Yoneyama H, Kurosawa Y, et al (1998) Do intronic mutations affecting splicing of WT1 exon 9 cause Frasier syndrome? *J Med Genet* 35: 45–48

33 Demmer L, Primack W, Loik V, Brown R, Therville N and McElreavey K (1999) Frasier syndrome: a cause of focal segmental glomerulosclerosis in a 46,XX female. *J Am Soc Nephrol* 10: 2215–2218

34 Denamur E, Bocquet N, Mougenot B, Da Silva F, Martinat L, Loirat C et al (1999) Mother to child transmitted WT1 splice-site mutation is responsible for distinct glomerular diseases. *J Am Soc Nephrol* 10: 2219–2223

35 Lala DS, Rice DA and Parker KL (1992) Steroidogenic factor I, a key regulator of steroidogenic enzyme expression, is the mouse homolog of fushi tarazu-factor I. *Mol Endocrinol* 6: 1249–1258

36 Honda S-I, Morohashi K-I, Nomura M, Takeya H, Kitajima M and Omura T (1993) Ad4BP regulating steroidogenic P-450 gene is a member of steroid hormone receptor superfamily. *J Biol Chem* 268: 7494–7502

37 Evans RM (1988) The steroid and thyroid hormone receptor superfamily. *Science* 240: 889–895

38 Parker KL and Schimmer BP (1997) Steroidogenic factor 1: a key determinant of endocrine development and function. *Endocrine Rev* 18: 361–377

39 Shen WH, Moore CCD, Ikeda Y, Parker KL and Ingraham HA (1994) Nuclear receptor steroidogenic factor 1 regulates MIS gene expression: a link to the sex determination cascade. *Cell* 77: 651–661

40 Hatano O, Takayama K, Imai T, Waterman MR, Takakusu A, Omura T et al (1995) Sex-dependent expression of a transcription factor, Ad4BP, regulating steroidogenic P-450 genes in the gonads during prenatal and postnatal rat development. *Development* 120: 2787–2797

41 Giuili G, Shen WH and Ingraham HA (1997) The nuclear receptor SF-1 mediates sexually dimorphic expression of Mullerian Inhibiting Substance, *in vivo*. *Development* 124: 1799–1807

42 Luo X, Ikeda Y and Parker KL (1994) A cell-specific nuclear receptor is essential for adrenal and gonadal development and for male sexual differentiation. *Cell* 77: 481–490

43 Sadovsky Y, Crawford PA, Woodson KG, Polish JA, Clements MA, Tourtellotte LM et al (1995) Mice deficient in the orphan receptor steroidogenic factor 1 lack adrenal glands and gonads but express P450 side-chain-cleavage enzyme in the placenta and have normal embryonic serum levels of corticosteroids. *Proc Natl Acad Sci USA* 92: 10939–10943

44 Ingraham HA, Lala DS, Ikeda Y, Luo X, Shen WH, Nachtigal MW et al (1994) The nuclear receptor steroidogenic factor 1 acts at multiple levels of the reproductive axis. *Genes Dev* 8: 2302–2312

45 Shinoda K, Lei H, Yoshii H, Nomura M, Nagano M, Shiba H et al (1995) Developmental defects of the ventromedial hypothalamic nucleus and pituitary gonadotroph in the Ftz-F1-disrupted mice. *Dev Dyn* 204: 22–29

46 Ikeda Y, Luo X, Abbud R, Nilson JH and Parker KL (1995) The nuclear receptor steroidogenic factor 1 is essential for the formation of the ventromedial hypothalamic nucleus. *Mol Endocrinol* 9: 478–486

47 Canteras NS, Simerly RB and Swanson LW (1994) Organization of projections from the ventromedial hypothalamic nucleus of the hypothalamus: a *Phaseolus vulgaris*-leucoagglutinin study in the rat. *J Comp Neurol* 348: 41–79

48 Oba K, Yanase T, Nomura M, Morohashi K, Takayanagi R and Nawata H (1996) Structural characterization of human Ad4BP (SF-1) gene. *Biochem Biophys Res Commun* 226: 261–267

49 Wong M, Ramayya MS, Chrousos GP, Driggers PH and Parker KL (1996) Cloning and sequence analysis of the human gene encoding steroidogenic factor 1. *J Molec Endocrinol* 17: 139–147

50 Ramayya MS, Zhou J, Kino T, Segars JH, Bondy CA and Chrousos GP (1997) Steroidogenic factor 1 messenger ribonucleic acid expression in steroidogenic and nonsteroidogenic human tissues: Northern blot and *in situ* hybridization studies. *J Clin Endocrinol Metab* 82: 1799–1806

51 Achermann JC, Ito M, Ito M, Hindmarsh PC and Jameson JL (1999) A mutation in the gene encoding steroidogenic factor-1 causes XY sex reversal and adrenal failure in humans. *Nat Genet* 22: 125–126

52 Nachtigal MW, Hirokawa Y, Enjeart-VanHouten DL, Flanagan JN, Hammer GD and Ingraham HA (1998) Wilms' tumor 1 and Dax-1 modulate the orphan nuclear receptor SF-1 in sex-specific gene expression. *Cell* 93: 445–454

53 De Santa Barbara P, Bonneaud N, Boizet B, Desclozeaux M, Moniot B, Sudbeck P, et al (1998) Direct interaction of SRY-related protein SOX9 and steroidogenic factor 1 regulates transcription of the human anti-Müllerian hormone gene. *Mol Cell Biol* 18: 6653–6665

54 Arango NA, Lovell-Badge R and Behringer RR (1999) Targeted mutagenesis of the endogenous mouse Mis gene promoter: *in vivo* definition of genetic pathways of vertebrate sexual development. *Cell* 99: 409–419

55 Zanaria E, Muscatelli F, Bardoni B, Strom TM, Guioli S, Guo W et al (1994) An unusual member of the nuclear hormone receptor superfamily responsible for X-linked adrenal hypoplasia congenita. *Nature* 372: 635–641

56 Swain A, Zanaria E, Hacker A, Lovell-Badge R and Camerino G (1996) Mouse Dax-1 expression is consistent with a role in sex determination as well as in adrenal and hypothalamus function. *Nature Genet* 12: 404–409

57 Ikeda Y, Swain A, Weber TJ, Hentges KE, Zanaria E, Lalli E et al (1996) Steroidogenic factor 1 and Dax-1 colocalize in multiple cell lineages: potential links in endocrine development. *Mol Endocrinol* 10: 1261–1272

58 Ito M, Yu R and Jameson JL (1997) DAX-1 inhibits SF-1-mediated transactivation via a carboxy-terminal domain that is deleted in adrenal hypoplasia congenita. *Mol Cell Biol* 17: 1476–1483

59 Crawford PA, Dorn C, Sadovsky Y and Milbrandt J (1998) Nuclear receptor DAX-1 recruits nuclear receptor corepressor N-CoR to steroidogenic factor 1. *Mol Cell Biol* 18: 2949–5629

60 Zazopoulos E, Lalli E, Stocco DM and Sassone-Corsi P (1997) DNA binding and transcriptional repression by DAX-1 blocks steroidogenesis. *Nature* 390: 311–315

61 Burris TP, Guo W, Le T and McCabe ER (1995) Identification of a putative steroidogenic factor-1 response element in the DAX-1 promoter. *Biochem Biophys Res Commun* 214: 576–581

62 Yu RN, Ito M and Jameson JL (1998) The murine Dax-1 promoter is stimulated by
   SF-1 (steroidogenic factor-1) and inhibited by COUP-TF (chicken ovalbumin upstream promoter-transcription factor) via a composite nuclear receptor regulatory element. *Mol Endocrinol* 12: 1010–1022
63 Kim J, Prawitt D, Bardeesy N, Torban E, Vicaner C, Goodyer P et al (1998) The Wilms'
   tumor suppressor gene (wt1) product regulates Dax-1 gene expression during gonadal
   differentiation. *Mol Cell Biol* 19: 2289–2299

Genes and Mechanisms in Vertebrate Sex Determination
ed. by G. Scherer and M. Schmid
© 2001 Birkhäuser Verlag Basel/Switzerland

Sry, Sox9 and mammalian sex determination

Peter Koopman

Centre for Molecular and Cellular Biology, The University of Queensland, Brisbane, QLD 4072, Australia

Summary. Sry is the Y-chromosomal gene that acts as a trigger for male development in mammalian embryos. This gene encodes a high mobility group (HMG) box transcription factor that is known to bind to specific target sequences in DNA and to cause a bend in the chromatin. DNA bending appears to be part of the mechanism by which *Sry* influences transcription of genes downstream in a cascade of gene regulation leading to maleness, but the factors that co-operate with, and the direct targets of, *Sry* remain to be identified. One gene known to be downstream from *Sry* in this cascade is *Sox9*, which encodes a transcription factor related to *Sry* by the HMG box. Like *Sry*, mutations in *Sox9* disrupt male development, but unlike *Sry*, the role of *Sox9* is not limited to mammals. This review focuses on what is known about the two genes and their likely modes of action, and draws together recent data relating to how they might interconnect with the network of gene activity implicated in testis determination in mammals.

Discovery of *Sry*, the Y-linked testis determinant

It has been known for some decades that in mammals, maleness is determined by the Y chromosome. Cytogenetic analyses in mice and humans showed conclusively that the Y chromosome carries a genetically dominant locus that normally induces male development in XY individuals [1–3]. This locus was given the acronym *TDF* (*testis-determining factor*) in humans and *Tdy* (*testis-determining gene on the Y*) in mice, and is now known to harbour the gene *Sry* (denoted *SRY* in humans).

The molecular strategy used to narrow down the region of search and eventually isolate the gene *SRY* is now regarded as a classic success story of positional cloning. Most positional-cloning projects involve identification of a candidate gene by analysis of translocations and deletions that result in an abnormal phenotype. In the case of sex determination, this was most readily approached by studying sex reversal in humans. The majority of human XX males possess Y-derived DNA sequences, transferred to the paternal X chromosome by aberrant X-Y interchange during meiosis [4, 5]. When DNA from four such XX males was analyzed, all were positive for Y-specific markers located in the 35 kb immediately adjacent to the pseudoautosomal boundary [6]. This 35-kb region was searched for conserved Y-linked sequences, an open reading frame found, and the corresponding gene dubbed *SRY* (*sex-determining region Y gene*). A homologous gene was found on the mouse Y [7].

This gene had all the properties expected of *TDF*, as elaborated in the following sections. Importantly, mutations in *SRY* were identified in XY females, showing that *SRY* function is normally required for testis development [8, 9]. The role of *Sry* in testis development was confirmed in transgenic mouse experiments, in which XX mice bearing *Sry* were able to develop as males [10]. These experiments demonstrated that only one gene from the Y chromosome is necessary and sufficient to initiate the cascade of male development, and that *Sry* encodes the testis-determining factor.

Structure and function of SRY

The role of the HMG box: DNA binding and bending

Sequence analysis of *Sry* revealed a region encoding a 79-amino acid motif that has come to be known as an HMG box, due to its presence in some of the high-mobility-group class of nonhistone proteins that associate with DNA [11]. This observation placed SRY among a set of sequence-specific DNA-binding transcription factors that include TCF1 and LEF1 [12–14]. Human SRY protein was shown to bind specifically to the sequences AACAAAG and, preferentially, AACAAT, the concensus binding sites of TCF1 and LEF1, respectively [15–18]. Mutations associated with XY sex reversal in humans were found to impair the binding of SRY protein to these target sites in electrophoretic mobility shift assays [16], indicating that the ability to bind to DNA is an integral part of the biochemical mechanism by which SRY directs male development. These observations sat well with expectations that the testis-determining factor was a cell-autonomous regulatory molecule at the top of a cascade of gene regulation leading to testis formation.

Further study revealed that SRY protein from humans and mice is able to induce bending of DNA by angles of 60–85° [19–21]. Similar properties were reported for LEF1 [20], leading to the suggestion that these proteins may act as "architectural" transcription factors, which are thought to act by influencing chromatin structure in the regulatory regions of target genes, allowing the assembly of regulatory complexes [22, 23]. Elegant *in vitro* studies showed that the activity of LEF1 in regulating the *TCRα* enhancer depended critically on the relative placement of its binding site and those of two cofactors, Ets-1 and PEBP2α, within the enhancer [24], as would be expected if the above model is correct. *Sry* was able to partially replace *LEF1* in stimulating enhancer function in transfection assays. Mutant SRY proteins produced by some human XY females have been found to be defective not in DNA binding, but in their ability to bend DNA [25], in support of an architectural model of transcriptional regulation by SRY.

What does the remainder of the SRY protein tell us about its likely mode of action in sex determination? The non-HMG box sequence of SRY is remarkably poorly conserved between species (Fig. 1). This is partly explicable by the location of *Sry* on the Y chromosome, conducive to a

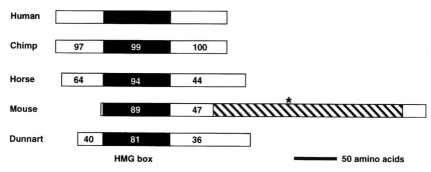

Figure 1. Sequence and topological comparisons of SRY proteins from various mammalian species. Linear schematic diagrams of predicted proteins are shown, with the HMG box in black. Numbers show amino acid identity to human SRY in the N-terminal, HMG box and C-terminal regions of the proteins, respectively. Sequences were sourced from GenBank, accession numbers L08063 (human), X86380 (*Pan troglodytes*, chimp), AB 004572 (*Equus caballus*, horse), U70655 (*Mus musculus musculus*, mouse), and S46279 (*Sminthopsis macroura*, stripe-faced dunnart). Identities were calculated using the University of Wisconson Genetics Computer Group program GAP, accessed through the Australian National Genome Information Service. The mouse protein contains a large, glutamine-rich repetitive region shown as a cross-hatched box. Mouse SRY in the subspecies *M. m. domesticus* has a premature stop codon that truncates this domain at the position indicated by an asterisk.

greater degree of sequence drift than that seen in autosomal or X-linked genes. Even so, the frequency of nonconservative substitutions in SRY is far greater than expected, to the extent that some workers have postulated positive selection for sequence change in *Sry* during evolution [26, 27]. Not only the primary structure but also the general topology of SRY can vary dramatically between species, in terms of the sizes of the regions N- and C-terminal to the HMG domain (Fig. 1). Further, mutations in human *SRY* that result in XY sex reversal are almost exclusively found in the HMG box (Fig. 2). Taken together, these observations point to a reliance of SRY largely on the HMG box for function, conferring at least in part an architectural mode of action.

Nuclear localization and phosphorylation

Given that the HMG box of SRY is responsible for DNA binding, can other essential features of a transcription factor be ascribed to SRY? One of these features is the ability of the protein to be transported to the nucleus after assembly in the cell cytoplasm. Recently, two independent nuclear localization signals have been identified at opposite ends of the HMG box of SRY [28, 29]. Some human XY females carry mutations in these nuclear localization signals (Fig. 2). In one case (R62G), transfection of a mutant protein expression construct into COS-7 cells revealed no differences in nuclear localization compared with wild-type constructs [29]. It remains to

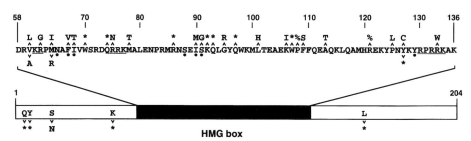

Figure 2. Sex-reversing mutations in human SRY leading to gonadal dysgenesis or hermaphroditism. A linear schematic diagram of the human SRY protein is shown, with the HMG box
in black. The 79-amino acid HMG box sequence is shown above in single letter code. Missense
mutations are denoted by the substituted amino acid above or below the normal sequence, nonsense mutations by "*" and frameshifts by "%". Note that the vast majority of these mutations
affect the HMG box. The effects of many of these on the structure and function of the HMG box
have been described [16, 25, 77, 181]. Amino acids involved in DNA contact are indicated with
a dot [181]. The bipartite and basic cluster nuclear localization signals [28, 29] are underlined.
Mutation data are from literature current to December 1999 [8, 9, 39, 77, 79, 182–201].

be seen whether other mutations do impede nuclear transport, implicating
a mode of sex reversal additional to reduced DNA binding or bending by
the HMG box.

Phosphorylation is one of the major mechanisms by which the activity of
transcription factors can be modulated in the cell. Recently, the cyclic
AMP-dependent protein kinase (PKA) has been found to phosphorylate
human SRY protein on serine residues located in the N-terminal part of the
protein [30]. This phosphorylation event was shown to positively regulate
SRY DNA-binding activity, providing evidence for posttranslational modulation of SRY activity during sex determination. Curiously, the PKA phosphorylation site (RRSSS) is conserved among primates but not other
mammalian orders.

Transcriptional activator or repressor?

While the conventional view is that SRY is likely to set in train a pathway
of gene regulation leading to testis determination and differentiation by
activating one or more downstream genes, an alternative suggestion is that
SRY may act as a transcriptional repressor. In this latter model, SRY may
act to repress genes that activate the female pathway of development, or to
repress a repressor of the male pathway. The repressor model is supported
by genetic studies of XX maleness. The frequency of XX males in the
human population is about 1 in 20,000 [31], and it is estimated that 10% of
these cases are *SRY*-negative [32, 33], implying that the frequency of *SRY*-
negative XX maleness is around 1 in 200,000. *Sry*-negative XX sex reversal
appears to be even more common in dogs [34]. This frequency is generally agreed to be too high to be explained by sporadic gain-of-function muta-

tions in male-determining genes downstream from *SRY*, and is perhaps more likely to result from loss-of-function of a repressor of maleness downstream from *SRY* [35]. *In vitro* evidence that SRY can act as a conventional repressor of transcription has also been presented [30]. However, no transcriptional repressor of maleness has been identified in the male sex-determining pathway, nor is there any *a priori* reason why such a repressor needs to be a direct target of SRY. An alternative model is that SRY protein could compete with, or otherwise block, the binding of another regulatory factor, or may impede rather than facilitate the assembly of a transcriptional complex through its bending properties.

Interaction of SRY with other proteins

A further essential requirement is that SRY should possess a mechanism by which it can specifically act on its target genes. SRY is a member of a large family of transcription factors, known as SOX proteins, related by similar HMG box sequences. One of these, SOX9, will be discussed in some detail below. All SOX proteins characterized to date can bind to similar target sequences, and functional specificity is presumably achieved through interaction with distinct cofactors in different regulatory contexts [36, 37]. Two interacting factors have so far been described for SRY. One of these is a nuclear factor, SIP1, containing PDZ protein interaction domains [38]. SIP1 was found to interact not with the HMG box, but instead with the C-terminal seven amino acids of human SRY. These amino acids are conserved in chimpanzee and gorilla SRY, but not in gibbon, orangutan or any of the other mammalian SRY proteins characterized to date. A base substitution generating a premature stop codon C-terminal to the HMG box in human SRY and resulting in XY gonadal dysgenesis has been described [39]. Such a mutation would affect the putative SIP1 interaction motif, providing a possible explanation for the phenotype, but may also affect other motifs or alter protein stability. In a separate study, Harley and colleagues found that the HMG box of SRY is a calmodulin binding domain [40]. Calmodulin interacts with a nuclear localization signal, and may therefore modulate nuclear import of SRY ([40] and V. Harley, personal communication).

Despite the overwhelming evidence regarding the importance of the HMG box for SRY function, a second, highly conspicuous domain is found in mouse SRY (Fig. 1). This domain in the *Mus musculus molossinus*-derived Y chromosomes, found in common laboratory strains such as 129, occupies over half the SRY protein (223 of 395 amino acids) at its C-terminus, is encoded mainly by a CAG trinucleotide repeat and is rich in glutamine and histidine residues [41]. This sequence is arranged as 19 blocks of 2–13 glutamine residues interspersed by a conserved, highly polar spacer of sequence FHDHH or similar. This domain is able to function as a transcriptional activator in a GAL4 assay in cultured cells [42]. Balanced against this observation is the absence of this domain in all other genera, and the finding that no substitute *trans*-activation domain exists in human SRY [42], suggesting that the glutamine-rich region does not play an activation role, nor indeed any role, *in vivo*. Even within the species *M. musculus*, this domain is variable: a stop codon truncates the glutamine-rich repeat region less than halfway through in *domesticus* subspecies [43] (Fig. 1). This truncated domain is unable to activate transcription *in vitro* [44]. We have tested directly whether the glutamine-rich repeat domain of mouse SRY is functionally relevant, using a transgenic mouse assay [45]. Constructs lacking this domain were unable to induce sex reversal, suggesting that the glutamine-rich repeat domain is essential for mouse SRY function, and that mouse SRY protein differs in its biochemical mode of action from SRY in other species. In support of this conclusion, a protein that specifically interacts with the glutamine-rich repeat region of mouse SRY has recently been reported, but the nature of this protein and its mode of action have not been determined [46, 47].

In summary, while it appears that SRY protein acts *via* its HMG domain to influence the transcription of other genes, basic information pertaining to what this influence is and how it is brought about is still lacking. Clearly target genes need to be identified as a priority in further clarifying the mode of action of SRY as a transcription factor.

Expression of *Sry*

Temporal and spatial profile of Sry transcription

In order to bring about male sex determination, *Sry* needs to be active in the gonadal primordia, known as the genital ridges, of the XY embryo. Shortly after the cloning of *Sry*, it was established that this gene is indeed expressed in the somatic cells of the genital ridges in mouse embryos [48]. This expression begins soon after the genital ridges first arise, about 10.5 days *post coitum* (dpc), reaches a peak around 11.5 dpc and is maintained only until the first morphological signs of testis differentiation become apparent at about 12.5 dpc [48–50]. Expression in mice was found to be specific to the genital ridges [51], although low-level expression in mouse, bovine and human blastocyst-stage embryos has been noted [52–54]. This expression profile is compatible with SRY acting specifically in the genital ridges to trigger a pathway of gene expression leading to testis development, with no continuing requirement for SRY expression in the maintenance of that pathway.

Once again, mice may not be truly representative of all mammals in this regard. A wide variety of expression profiles have been noted among different mammalian species. In humans, marsupials and sheep, *Sry* transcription appears to be much less tissue- and stage-specific than in mice. For example, the timing of *Sry* transcription in sheep and Tammar wallabies encompasses that of testis differentiation by a broad window of several days either side [55, 56]. In humans and wallabies, expression in several fetal and adult tissues has been noted [55, 57]. These observations suggest either that *Sry* has roles other than in sex determination in these species, or that expression outside of the genital ridge or beyond the narrow time window of sex determination has no functional relevance. If the former explanation is true, the effects of *Sry* in nongonadal tissues must be subtle, as mutation and indeed absence of *Sry* appears only to affect the testes and processes depending on the hormonal output of the testes. It may well be that the relatively extensive expression of *Sry* in nonmurine species is redundant, and that mice have retained only the minimal expression profile required for male sex determination.

Consistent with this view is another curious quirk of *Sry* expression in mice. In the course of attempts to isolate a complementary DNA (cDNA) for *Sry*, libraries were screened from the most abundant source of *Sry* expression, the adult testis. cDNA clones were repeatedly generated in which sequences normally located 3′ to the HMG box were found in a 5′ position, abutting sequences upstream from the HMG box. These results were confirmed by RNase protection and rapid amplification of cDNA ends-polymerase chain reaction (RACE-PCR). This apparent conundrum was solved by the demonstration that a circular transcript is produced in adult testes [58]. Circularization is thought to result from the unique structure of the *Sry* locus: mouse *Sry* is embedded in a large inverted repeat [41]. The adult testis transcript is believed to initiate within one arm of this repeat [59], encompass the open reading frame and terminate in the other arm of repeated sequence. This scenario would generate a stem-loop transcript, and a splice involving donor and acceptor sequences present in the loop could generate the final circle [58, 60]. Why this occurs in mice is not known. The circular transcript is not translated and therefore appears to be redundant, in contrast to the functional transcript produced in the genital ridges, and may be used as a means of suppressing *Sry* function in sites of gene expression other than the fetal gonads. However, evidence from transgenic studies in my laboratory indicates that loss of tissue specificity of *Sry* expression is not detrimental in mice (J. Bowles and P. Koopman, unpublished analysis).

Cellular consequences of Sry expression

The primordial gonad is composed of a number of distinct lineages, each of which has the potential to differentiate into testicular or ovarian counterparts, depending on signals received. The primordial germ cells can either mitotically arrest and subsequently develop into prospermatogonia, or enter meiosis and become oogonia [61]. This decision is a result rather than an effector of sex determination, as embryos lacking germ cells are able to undergo somatically normal sex determination [62, 63]. Clearly germ cells cannot be the site of *Sry*'s action, and indeed mouse mutant embryos lacking germ cells retain *Sry* expression [48]. Somatic cell lineages in the developing gonads include supporting cell precursors that can develop into Sertoli cells in males or follicle cells in females, steroidogenic precursor cells that go on to become either Leydig cells or theca cells, and mesenchymal cells that can either contribute to the peritubular myoid cells and vasculature of the testis, or organize into ovarian connective tissue and vasculature. Genetic studies involving XX ↔ XY chimeric mice have shown that of these cell types, *TDF/Tdy* expression is required in a cell-autonomous fashion only in the supporting cell lineage [64, 65]. These studies were supported by direct observation of *Sry* expression in Sertoli cell precursors [66]. These results imply that the role of SRY is to act as a switch that influences the differentiation of Sertoli cells, and that factors produced by differentiating Sertoli cells influence the differentiation and organization of the remaining cell lineages.

Elegant organ culture and genetic studies by Capel and colleagues have provided intriguing insights into the part played by *Sry* in cellular events in testis development. It has been known for some time that cells migrate from the mesonephros into the developing testis, and that these cells contribute to the interstitial cell population [67]. If this migration is blocked, testis differentiation cannot proceed. Using wild-type and genetically marked ROSA26 transgenic mouse tissue in *in vitro* recombination experiments, Capel and colleagues showed that this migration is specific for XY gonads [68]. It has now been determined that this effect is dependent on *Sry*, rather than another Y-linked gene or the presence of one vs. two X chromosomes, since it occurs in XX gonads of embryos transgenic for *Sry* [69]. This migration is capable of inducing cord formation even in pieces of XX genital ridge placed between mesonephros and XY genital ridge, demonstrating the critical role of mesonephric cell migration in the formation of testis cords [70]. These studies indicate that an important function of *Sry* is to stimulate production of a diffusible chemoattractant molecule that signals to mesonephric cells and induces their migration and subsequent cord formation in the testes.

Importance of expression levels for SRY function

Many lines of evidence combine to indicate that expression levels of *Sry* are important for sex determination. In humans, different degrees of masculinization can be seen in XX siblings inheriting the same Y chromosome-derived fragment after aberrant meiotic X-Y interchange in their father's germ line [71]. This variation must be due either to different degrees of mosaicism for cells inactivating the X chromosome bearing *SRY* in the Sertoli cell lineage, or differences in *SRY* expression levels due to differential spreading of X chromosome inactivation, or both. In mice, XY sex reversal is associated with two types of mutation that appear to exert their effects through suppressing *Sry* expression levels. Capel and colleagues noted that deletion of repeat sequences at some distance proximal to *Sry* on the mouse Y short arm result in reduced expression and sex reversal [72]. More recently, the sex reversal seen when crossing an *M. m. musculus* strain *Poschiavinus* Y chromosome (Y^POS) onto an *M. m. domesticus* background (in particular C57BL6) [73] has been ascribed to reduced *Sry* expression levels from Y^POS. In an elegant study, Nagamine and colleagues showed by quantitative reverse transcriptase-polymerase chain reaction (RT-PCR) that levels of *Sry* expression from different mouse Y chromosomes correlated with the degree to which they cause sex reversal on a *domesticus* background [74]. It appears that a critical threshold must be achieved by a certain stage in genital ridge development for the supporting cell population to be pushed towards Sertoli cell differentiation.

Regulation of Sry transcription

While expression levels of *Sry* are clearly important, it is not yet known how *Sry* is regulated. Experimental attempts to define the regulatory regions of *Sry* have been hampered by poor conservation of upstream sequences, low expression levels, lack of suitable cell lines to use in transfection studies and unsuitability of β-galactosidase as a reporter molecule for *Sry* expression in transgenic mice. Comparison of upstream sequences from closely related species such as primates has identified some conserved elements that may be relevant to *Sry* regulation [75, 76]. Mutation searches in *SRY* upstream sequences in cases of XY sex reversal have typically found no sequence changes [77–80]. Poulat and coworkers [80] found one point mutation at position −75 in an XY female, but the significance of this mutation is unclear, as the father's DNA could not be analyzed. Certainly, deletion of 33–60 kb of upstream sequence, starting 1.8 kb 5′ to *SRY*, causes XY sex reversal [81]. Downstream sequences may also be implicated in *SRY* regulation in humans [79, 82]. Whatever the eventual outcome of *Sry* regulatory studies, important *cis*-sequences and their *trans*-acting regulatory factors identified in one species such as mice may or may not be broadly relevant, given the differences in *Sry* structure and expression between species. Human *SRY* is unable to cause male sex determination in transgenic mice [10], and it is not yet known whether this is due to structural or regulatory incompatibilities between the human gene or protein and the host mouse cells. Clearly, however, investigations relating to how the *Sry* gene is regulated pose a formidable challenge.

Evolution of *Sry*

What is the origin of *Sry* and how did it come to control a decision of such fundamental importance for the individual and for the species? Obviously it is possible only to speculate on these questions, based on available evidence. First, *Sry* is found only in mammals. Extensive searches among other vertebrate classes for genuine orthologues of *Sry* (that is, genes containing an HMG box that are present in one sex but not the other) have yielded only *Sox* genes that are not sex-specific [83, 84]. Many eutherian and a few marsupial and monotreme species have been studied, and *Sry* has been found in all of these bar three, namely two vole species *Ellobius lutescens* and *E. tancrei* [85], and the spiny rat *Tokudaia osimensis* [86]. Assuming that these species have diverged on their own evolutionary tangents, it seems that the origin of *Sry* is inextricably linked to the genesis of the class Mammalia.

The X and Y chromosomes are thought to have arisen as a pair of autosomes that began to diverge when the Y took on a male-determining function. *Sry* may thus have begun its life as a *Sox* gene on the ancestral autosome. The X chromosome of humans, mice and marsupials carries a gene, *Sox3*, which may be the gene from which *Sry* evolved [87–89]. *Sox3* is primarily expressed in the developing central nervous system, although transcripts have been reported in the genital ridges of both sexes in mice and chickens [89–93]. It is tempting to speculate that *Sox3* may once have been, or may still be, involved in female sex determination, and that competition between SRY and SOX3 proteins was or is the basis of a male-female dichotomy [94]. There is currently no evidence for a role for *Sox3* in determination of either sex [87, 95]. However, it is not necessary for *Sox3* to have such a role in order to have been the ancestor of *Sry*, as the latter may have taken on an entirely novel function. Further, given the rate of evolution of *Sry*, it is impossible to say with certainty which *Sox* gene *Sry* most resembles. It is not yet known whether other *Sox* genes reside on the X chromosome and can be considered as possible ancestors of *Sry*.

Relationship of *Sry* to other genes implicated in sex determination

A number of genes have been identified that have a critical role in the male or female sex-determining pathway, and may therefore interact in some way with *Sry*. These are described briefly below, and three – *DAX1*, *SF1* and *WT1* – are discussed in detail elsewhere in this volume. A schematic diagram of the cellular and molecular interactions during sex determination and gonadal development is shown in Figure 3. These interrelationships have been reviewed elsewhere by Swain and Lovell-Badge [96].

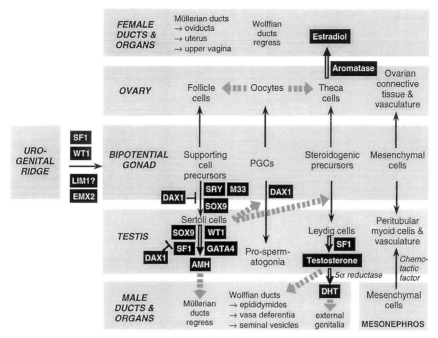

Figure 3. Cellular and molecular interactions during gonadal induction. Pathways of cellular differentiation and/or migration are indicated by black arrows, biosynthetic pathways by hollow arrows, and hormonal or unknown signalling pathways by large dotted arrows. Effector genes or gene products are shown in black boxes. DHT, dihydrotestosterone.

Amh

Anti-Müllerian hormone (AMH, otherwise known as Müllerian-inhibiting substance or MIS) causes regression of the Müllerian (female) duct system. Because it is the first identifiable Sertoli cell product and appears soon after the onset of *Sry* expression, it was for some time suspected that the *Amh* gene is directly regulated by SRY. However, it now appears that at least some intermediaries such as SOX9 and SF1 are involved (see below). In any case, it is known that AMH is not an integral link in the male-determination pathway, as mice lacking AMH show normal testis development [97].

SF1

The orphan nuclear receptor steroidogenic factor 1 (SF1), otherwise known as Ad4BP, is a key component of the pathway of gonadal and adrenal development, and is known to regulate male steroid biosynthesis. *Sf1* transcripts are present in the gonads of both sexes in mice between 9 and 12 dpc, whereafter the gene is downregulated in ovaries [98, 99]. Mice homozygous for a null allele in *Sf1* show normal early development of the genital ridges, but a complete block in subsequent gonadal development in both sexes, among other defects [100]. A mutation in human *SF1* has been associated with XY sex reversal and adrenal failure [101]. The *Amh* promoter contains an SF1 binding site which is critical for expression of the gene in Sertoli cells *in vitro* and in transgenic *Amh*-reporter mice [102, 103]; mutation of this site *in vivo* results in decreased levels of *Amh* expression [104]. Together, these data suggest that SF1 has at least three important roles in sexual development, one early in the establishment of the gonad prior to the expression of *Sry*, another later in male differentiation, regulating *Amh*, and another in regulating steroid synthesis in Leydig and theca cells.

WT1

Targeted disruption of the Wilms' tumour-associated zinc-finger gene *WT1* also results in a blockage of gonad development in mice [105]. As with *Sf1* deficiency, the initial stages of genital ridge formation are unaffected, suggesting that *SF1* and *WT1* play important roles in the maturation of the genital ridges, rather than sex determination *per se*, perhaps by establishing an environment in which *Sry* can act to commit cells to the male fate. In humans, dominant-negative *WT1* mutations are associated with XY pseudo-hermaphroditism in Denys-Drash syndrome [106]. It has been suggested that WT1 may act as a regulator of *Sry* expression [107], and potential WT1 binding sites are present upstream of both mouse and human *Sry*. However, direct evidence for this regulation is lacking.

WT1 does not appear to be required for *Sf1* expression, as transcripts of the latter are detectable in *Wt1* knockout mice, but may contribute to *Sf1* upregulation [108]. Evidence has also been presented that WT1 can activate the *Dax1* promoter through direct binding to GC-rich sequences near the TATA box [109].

Emx2

The homeobox-containing transcription factor gene *Emx2* is expressed in genital ridges as well as the Wolffian ducts, mesonephric tubules and coelomic epithelia during mouse embryo development. This expression pattern is reminiscent of that of *Wt1*. Mice deficient for *Emx2* show impaired gonadal and kidney development [110]. Mutant embryos have poorly developed gonads and the mesonephric tubules and Wolffian ducts degenerate. The role of *Emx2* in gonad development is not clear. *Wt1* expression is unaffected in the *Emx2* mutants, suggesting that *Emx2* acts downstream of *Wt1*.

LIM1

LIM1 is a homeobox gene that regulates production of an organizing molecule in many experimentally studied species. Targeted disruption of *Lim1* in mice resulted in complete absence of head structures and early lethality, but surviving fetuses lack kidneys and gonads [111]. It is widely assumed that *LIM1*, like *WT1* and *SF1*, is involved in maturation of the genital ridges, but its precise role in gonadal induction has not been analyzed in detail, due in part to the lethality of *Lim1* knockout embryos.

DAX1

Another orphan nuclear receptor gene, *DAX1* maps to a region of the human X chromosome which, when duplicated, causes male-to-female sex reversal [112–114], and it has been suggested that *Dax1* is an ovarian-determining gene [112]. Deletions involving *DAX1* do not disrupt testis differentiation, and *Dax1* expression in mice is downregulated with testis differentiation, but persists in the developing ovary [115]. Transgenic mice overexpressing *Dax1* have been shown to undergo male-to-female sex reversal, suggesting that *Dax1* and *Sry* act antagonistically [116]. However, targeted inactivation of *Dax1* in mice does not affect ovarian development, instead blocking spermatogenesis in males [117]. It is concluded that *Dax1* is a spermatogenesis and anti-maleness gene rather than an ovarian-determining gene [116, 118]. This conclusion is compatible with molecular observations that DAX1 protein can act as a transcriptional repressor of steroidogenesis [119–122].

Aromatase

In vitro binding of SRY to the *aromatase* promoter has also been reported [123]. Aromatase is a member of the cytochrome P-450 family and is responsible for the conversion of testosterone to estradiol. It is considered essential for male sex determination that the gene encoding aromatase be tightly repressed during testis development, and Haqq et al. have suggested that SRY is directly responsible for this repression. This is unlikely to be the case *in vivo*; *SRY* expression occurs in Sertoli cells, yet aromatase is a product of the steroidogenic lineage (Fig. 3).

Fra1

The *Fos*-related antigen gene *Fra1* has been implicated as a target of SRY, since transfection of a plasmid expressing human SRY protein significantly enhanced transcription of a *Fra1* reporter construct [124]. However, *Fra1* is not expressed in the gonads at or around the critical period of sex determination when *Sry* transcripts are present in mice, excluding a role for *Fra1* in sex determination and differentiation [50].

M33

XY homozygous null mutants for the polycomb-related homeobox gene *M33* develop as females or hermaphrodites [125]. *Sry* is apparently expressed normally in these mice although *Sox9* expression is reduced from levels normally seen in XY genital ridges (Y. Katoh-Fukui, unpublished observations). These data suggest that *Sry* does not lie downstream of *M33*, and that *M33* may act as an intermediate between *Sry* and *Sox9*, or may act co-operatively with *Sry* to directly activate or repress downstream genes.

GATA4

The zinc finger transcription factor GATA4 is expressed in the somatic cell lineages but not germ cells of genital ridges of both sexes as early as 11.5 dpc. The expression later becomes sexually dimorphic with protein detected exclusively in nuclei of newly differentiated Sertoli cells and expression downregulated in ovary shortly after its histological differentiation on day 13.5 [126]. The involvement of GATA4 in gonadal development and sexual dimorphism is supported by evidence that it may directly activate *Amh* transcription, apparently in concert with SF1 [126, 127].

Dmrt1

A vertebrate gene, *Dmrt1*, has been isolated which shows homology to the male sexual regulatory gene *mab-3* from *Caenorhabditis elegans* and the *Drosophila* sexual regulatory gene *doublesex* (*dsx*) [128]. In humans, *DMRT1* maps to the distal short arm of chromosome 9, a region associated with deletions that result in XY sex reversal. However, since a paralogous gene, *DMRT2*, as well as other unrelated genes, map to this interval [129], it is not possible to assess the role of *DMRT1* in sex reversal from the available patient data. Interestingly, *Dmrt1* maps to the Z chromosome in chickens, suggesting a possible dosage model for male sex determination in birds [130]. *Dmrt1* has been shown to be expressed specifically in developing gonads in mouse, chick and alligator embryos in both sexes before gonadal differentiation [131], then becoming up-regulated in testes and down-regulated in ovaries [131, 132], consistent with an early role in sex determination in vertebrates.

Sox9

As mentioned earlier in this review, *Sry* is just one of a family of genes related by the HMG box. These genes have come to be known as *Sox* (*SRY-related HMG box*) genes (for review, see [133]). A rule of thumb for defining *Sox* genes, as distinct from other HMG box genes, is that they are more than 50% identical to SRY at the amino acid level in the HMG box region. However, this cannot be regarded as strictly accurate, as several exceptions to this rule exist [134] (P. Koopman, unpublished analysis). *Sox* genes have been found in a wide variety of species representing insects, amphibians, ascidians, reptiles, birds, fishes and mammals. They are best characterized in mice, with nearly 30 *Sox* genes identified in whole or part in that species. Expression studies (see [133, 135] and references therein), knockout mouse experiments [136–138] and the identification of mutations in a number of human diseases [8, 9, 139–141) indicate that *Sox* genes are important for the differentiation and/or function of a number of cell types during embryonic development.

One *Sox* gene of special relevance for sex determination is *Sox9*. Expression of *Sox9* was first noted at sites of chondrogenesis in mouse embryos [142]. In humans, heterozygous defects in *SOX9* have been associated with the bone dysmorphology syndrome campomelic dysplasia (CD) [139, 140]. Curiously, a large proportion of CD patients show XY sex reversal [143, 144], demonstrating an important role for *Sox9* not only in skeletal development but also in sex determination. *Sox9* is one of only a few genes apart from *Sry* for which mutations have been shown to interfere with male sex determination.

Structure and function of SOX9

The HMG box: DNA binding and nuclear localization

The structure of SOX9 is that of a typical transcription factor with discrete DNA-binding and transcriptional *trans*-activation domains. The HMG box region of SOX9 has been shown to bind to the sequences AACAAT and AACAAAG [36, 145], typical of SRY and other SOX proteins. In addition, SOX9 has been found to bind to the variant sequences ATGAAT and CACAAT, found in the chondrocyte-specific enhancer in the first intron of the human type II collagen gene *COL2A1* [145, 146]. Mutation of these binding motifs abolishes chondrocyte-specific expression of a *COL2A1* reporter in transgenic mice [145], demonstrating not only the importance of SOX9 binding for *COL2A1* expression, but also that the canonical SOX binding site (WWCAAWG, where W = A or T) defined by binding site selection experiments *in vitro* is not necessarily used *in vivo*.

The importance of DNA binding for SOX9 function is reflected by the presence of mutations affecting the SOX9 HMG box in sex-reversed (XY female) CD patients [139, 140, 147–149] (Fig. 4). Impaired DNA-binding affinity has been demonstrated in a number of these cases [148]. It is likely that SOX9 binding results in chromatin bending, but bending brought about by normal SOX9, let alone impaired bending resulting from sex-reversing mutations in the HMG box, has not been investigated.

The HMG box of SOX9, like that of SRY, contains two independent nuclear localization signals [29] (Fig. 4). In mice, SOX9 protein appears predominantly cytoplasmic in the genital ridges of both sexes prior to 11.5 dpc, but moves to the nucleus at later stages of testis development [150], and it has been suggested that a factor or factors that interact with the nuclear localization signal(s) may mediate this change in subcellular localization [29]. No direct support for this theory has emerged, and mutations specifically affecting nuclear localization sequences of SOX9 have not yet been detected among campomelic dysplasia patients.

Transcriptional trans-activation

Sex-reversing mutations in *Sox9* are found throughout the gene, with no correlation between the type of mutation and the phenotype [148] (Fig. 4). Clearly, functional domains outside of the HMG box are implicated by these mutations. *In vitro* GAL4 activation assays have identified the C-terminal 108 amino acids in human SOX9 as a *trans*-activation domain

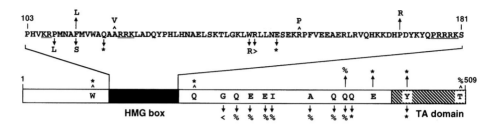

↑ CD/XY ♂   ^ CD/XX ♀   ↓ CD/XY ♀

Figure 4. Mutations in human *SOX9* associated with campomelic dysplasia and XY sex reversal. Linear schematic diagram of the 509-amino acid human SOX9 protein. The HMG box and
trans-activation (TA) domain are indicated, with the 79-amino acid HMG box sequence shown
above in single letter code. Missense mutations are denoted by the substituted amino acid,
whereas nonsense mutations are denoted by "*" and frameshifts by "%". Mutations affecting
splice donor (<) and acceptor (>) sequences have also been reported. Mutations designated
by an upward arrow do not affect sex determination (i.e. they are present in XY males). Those
designated by a downward arrow result in complete or partial XY sex reversal (XY females).
The effect on male sex determination of the mutations designated is not known (XX females).
The bipartite and basic cluster nuclear localization signals [29] are underlined. Mutation data
are from literature current to December 1999 [139, 140, 147–149, 202, 203].

[151]. In mice, the C-terminal 83 amino acids are capable of *trans*-activa-
tion [36]. The difference is likely due to the serendipitous choice of gene
fragments assayed in the respective laboratories rather than to any genuine
differences between the mouse and human proteins, and it may be that the
real size of the *trans*-activation domain is smaller still. In support of this
possibility, the C-terminal 43 amino acids of the SOX9 protein are particularly well conserved between the human, mouse, chicken and rainbow
trout sequences (Fig. 5). Whatever the actual size of the *trans*-activation-
domain, all of the CD mutations C-terminal to the HMG box are termi-
nation or frameshift mutations that would debilitate the *trans*-activation
ability of the protein. It would appear then that SOX9 is able to function as
a conventional transcription factor, but whether the protein has additional
or alternative architectural roles remains to be seen.

Other domains of SOX9

Unlike the situation in SRY, SOX9 shows a very high degree of sequence
conservation throughout the protein between species. This conservation is
typical of *Sox* genes other than *Sry*, and is particularly high for the HMG
box and *trans*-activation domains (Fig. 5), as might be expected. Human
and mouse SOX9 contain a conspicuous region of some 40 amino acids
composed entirely of proline (P), glutamine (Q) and alanine (A) residues
(Fig. 5). This region has been found to augment the potency of the C-
terminal *trans*-activation domain, but is unable to activate transcription

```
            10        20        30        40        50        60
             .         .         .         .         .         .
Human   MNLLDPFMKMTDEQEKGLSGAPSPTMSEDSAGSPCPSGSGSDTENTRPQENTFPKGEPD-
Mouse   MNLLDPFMKMTDEQEKGLSGAPSPTMSEDSAGSPCPSGSGSDTENTRPQENTFPKGDPD-
Chick   MNLLDPFMKMTEEQDKCISDAPSPTMSDDSAGSPCPSGSGSDTENTRPQENTFPKGEPD-
Trout   MNLLDPFLKMTDEQEKCFSDAPSPSMSEDSVGSPCPSGSGSDTENTRPSDNHLLLG-PDG
        ••••••••.•••.••.•  •  ••••.••.••  •••••••••••••••  .•   •  ••

Human   ---LKKESEEDKFPVCIREAVSQVLKGYDWTLVPMPVRVNGSSKNKPHVKRPMNAFMVWA
Mouse   ---LKKESEEDKFPVCIREAVSQVLKGYDWTLVPMPVRVNGSSKNKPHVKRPMNAFMVWA
Chick   ---LKKESDEDKFPVCIREAVSQVLKGYDWTLVPMPVRVNGSSKNKPHVKRPMNAFMVWA
Trout   VLGEFKKADQDKFPVCIRDAVSQVLKGYDWTLVPMPVRLNGSSKNKPHVKRPMNAFMVWA
          •  .•••••••••••••.••••••••••••••••••••.••••••••••••••••••••

Human   QAARRKLADQYPHLHNAELSKTLGKLWRLLNESEKRPFVEEAERLRVQHKKDHPDYKYQP
Mouse   QAARRKLADQYPHLHNAELSKTLGKLWRLLNESEKRPFVEEAERLRVQHKKDHPDYKYQP
Chick   QAARRKLADQYPHLHNAELSKTLGKLWRLLNESEKRPFVEEAERLRVQHKKDHPDYKYQP
Trout   QAARRKLADQYPHLHNAELSKTLGKLWRLLNEGEKRPFVEEAERLRVQHKKDHPDYKYQP
        •••••••••••••••••••••••••••••••.••••••••••••••••••••••••••••

Human   RRRKSVKNGQAEAEEEATEQTHISPNAIFKALQ-ADSPHSSSGMSEVHSPGEHSGQSQGPP
Mouse   RRRKSVKNGQAEAEEEATEQTHISPNAIFKALQ-ADSPHSSSGMSEVHSPGEHSGQSQGPP
Chick   RRRKSVKNGQSEQEEGSEQTHISPNAIFKALQ-ADSPQSSSSISEVHSPGEHSGQSQGPP
Trout   RRRKSVKNGQSEPEDG-EQTHISSGDIFKALQQADSP--ASSMGEVHSPSEHSGQSQGPP
        ••••••••••.•  ••• .  •••••• •• •    .•  .• •••• •••••••••••••

Human   TPPTTPKTDVQ-PGKADLKREGRPLPEGGRQPP-IDFRDVDIGELSSDVISNIETFDVNE
Mouse   TPPTTPKTDVQ-AGKVDLKREGRPLAEGGRQPP-IDFRDVDIGELSSDVISNIETFDVNE
Chick   TPPTTPKTDAQQPGKQDLKREGRPLAEGGRQPPHIDFRDVDIGELSSDVISNIETFDVNE
Trout   TPPTTPKTDLA-VGKADLKREGRPLQEGTGRQLNIDFRDVDIGELSSDVISNIEAFDVHE
        •••••••••  ••  •••••••••• •   .   ••••••••••••••••••••.•••.•

Human   FDQYLPPNGHPGVPATHGQVT-YTGSYGISS--TAATPASAGHVWMSKQQAPPPPPQQPP
Mouse   FDQYLPPNGHPGVPATHGQVT-YTGSYGISS--TAPTPATAGHVWMSKQQAPPPPPQQPP
Chick   FDQYLPPNGHPGVPATHGQVTTYSGTYGISS--SASSPAGAGHAWMAKQ--------QP-
Trout   FDQYLPPHGHPGMPGINGAQTSYTGSYRGISSNSIGQVGAGGHGWMSKQ-----------
        •••••••.•••••.••.•  .•  • •.•.•   •   •    ••  ••••.••

Human   QAPPAPQAPPQPQAAPPQQPAAPPQQPQAHTLTTLSSEPGQSQ--RTHIKTEQLSPSHYS
Mouse   QAPQAPQAPPQ-QQAPPQQPQAP-QQQQAHTLTTLSSEPGQSQ--RTHIKTEQLSPSHYR
Chick   -Q------PPQ----PPAQPPAQ------HTLPALSGEQGPAQQ-RPHIKTEQLSPSHYS
Trout   ----------Q------QQPISILS-----GGGGTGGEQGQSQGRTTQIKTEQLSPSHYS
                 •       ••  .            • •  .•     .•••••••••••

Human   EQQQHSPQQ-------IAYSPFNLPHYSP-SYPPITRSQYDYTDHQN-SSSYYSHAAGQG
Mouse   EQQQHSPQQ-------ISYSPFNLPHYRP-SYPPITRSEYDYADHQN-SGSYYSHAAGQG
Chick   EQQQHSPQQQQQQQQQLGYGSFNLQHYGS-SYPPITRSQYDYTEHQN-SGSYYSHAAGQS
Trout   E-QQGSPPQ------HVTYGSFNLQHYSASSYPSITRTQYDYSDHQGGANSYYSHAGAQG
        • •• •• •           . •  •••  ••  ••• •••,,•••,,••, .•••••• •

Human   TGLYSTFTYMNPAQRPMYTPIADTSGVPSIP-QTHSPQHWEQ-PVYTQLTRP
Mouse   SGLYSTFTYMNPAQRPMYTPIGDTSGVPSIP-QTHSPQDWEQ-PVYTQVTRP
Chick   GGLYSTFTYMNPTQRPMYTPIADTSGVPSIP-QTHSPQHWEQ-PVYTQLTRP
Trout   SGLYSFSSYMSPSQRPMYTPIADPTGVPSVPTQTHSPQHWEQQPVYTQLSRP
        ••••  ,•• •,•••••••• •  ,••••,• •••••• ••• •••••,,••
```

Figure 5. Comparison of SOX9 proteins from human, mouse, chicken and trout. Amino acids identical in all four sequences are indicated with a bold dot, and those subject to conservative change in some sequences with a smaller dot. The HMG box is underlined, and the C-terminal 83-amino acid region implicated in transcriptional *trans*-activation in the mouse is double-underlined. The pattern of conservation through the SOX9 protein in these diverse vertebrates suggests that regions other than the HMG box and *trans*-activation domains may be of functional significance. Data from [37, 139, 142, 204].

alone [152]. This sequence is highly degenerated in chicken *Sox9*, and is represented as only 5 residues (QQQQP) in trout (Fig. 5), and may therefore not be critical to SOX9 function.

Sequence comparisons also reveal several other very highly conserved regions that may serve as interfaces for interaction with other proteins. One might predict that SOX9 would interact with a variety of cofactors in a combinatorial fashion to achieve its different roles, presumably involving many different regulatory targets, during development [36, 37]. For example, SOX9 is known to directly activate *COL2A1* and the genes encoding the matrix components aggrecan and CD-RAP during chondrogenesis [145, 146, 153, 154]. Cooperation with a complex of SOX5 and SOX6 synergizes the activation of *COL2A1* [155], and other, as yet unidentified cofactors may be involved in *COL2A1* activation. Cofactors enhancing the effect of SOX9 on *aggrecan* and *CD-RAP* transcription are not yet known. In sex determination, SOX9 interacts with SF1 in activating *Amh* transcription [104, 156]. Clearly it will be of great interest to map the domains of SOX9 involved in these various liaisons.

Curiously, only some 75% of XY infants showing the skeletal abnormalities diagnostic for CD are sex-reversed [157]. In other words, some *SOX9* mutations result in CD with XY sex reversal and others in CD without XY sex reveral, whereas mutations seen in XX patients are uninformative with respect to their effects on male sex determination (Fig. 4). This may mean that the molecular cascade of chondrogenesis is more sensitive than that of sex determination to perturbations in SOX9 function. Alternatively, the majority of CD fetuses may develop ovotestes, which resolve to either testes or ovaries after birth. No cases of *SOX9* mutations resulting in sex reversal without CD have been found [148, 158], and the nature and distribution of the sex-reversing mutations in human *SOX9* (Fig. 4) do not point to any part of the protein as uniquely important for a role in sex determination.

Expression of *Sox9*

Much has been written about the expression of *Sox9* during chondrogenesis [36, 142, 159, 160], and it is beyond the scope of this review to reiterate these data. Accumulated data from several laboratories, culminating with analysis of the developmental potential of cells lacking *Sox9* [138], indicate that *Sox9* serves as an essential trigger for the differentiation of chondrocytes from mesenchymal cells in the embryo, analogous to the proposed role of *Sry* in stimulating Sertoli cell differentiation in mammals.

Expression of *Sox9* during sex determination suggests a role downstream from *Sry* in the Sertoli cell-differentiation pathway. This expression profile has been studied independently by two laboratories, with slightly different outcomes. Lovell-Badge and colleagues reported expression of *Sox9* in mouse genital ridges of both sexes at 10.5 dpc, whereas Koopman and colleagues did not detect *Sox9* transcripts in either sex at the same stage [150, 161]. The discrepancy is almost certainly due to differences in sensitivity in the two separate series of experiments. Both groups, however, reported a strong upregulation of *Sox9* expression in male genital ridges at 11.5 dpc. This appears to coincide with a downregulation of *Sox9* expression in the female. *Sox9* expression is subsequently maintained in testes during and after their differentiation. *Sox9* RNA expression is associated with developing testis cords, and is not due to germ cells, implicating Sertoli cells as the site of *Sox9* expression [161]. Sertoli cell-specific expression of SOX9 protein has been confirmed by immunohistochemistry [150, 162]. *Sox9* expression in Sertoli cells is not transient, but persists through to adulthood.

As with SRY, levels of functional SOX9 protein appear to be critical for male sex determination. Sex reversal associated with CD results from mutation of one copy of *SOX9*. As mentioned above, some of these mutations are likely to impair the ability of the SOX9 protein in an affected individual to bind to DNA. Other mutations would be expected to have no effect on DNA binding, but instead impair other biochemical functions of the protein such as transcriptional transactivation [151]. A third class of mutations is represented by translocation breakpoints upstream of *SOX9* [144, 163]; these would presumably reduce or abolish transcription from the mutant *SOX9* allele. A recent report of XX sex reversal associated with a duplication involving *SOX9* may further reflect the importance of correct levels of *SOX9* expression in sex determination [164].

Sry, Sox9 and the testis-determining pathway

How do *Sry* and *Sox9* interconnect with each other and with the other pieces of the sex-determination jigsaw identified to date? Several reviews have speculated on the nature of the mammalian sex-determining pathway [165–168]. The following discussion summarizes the current issues and pertinent observations relating to *Sry, Sox9* and their interrelationship in the context of sex determination. These relationships are schematically represented in Figure 3.

What regulates Sry?

WT1, SF1, EMX2 and LIM1 are implicated in the regulation of *Sry* expression by virtue of their early expression in the genital ridges, but it is also likely that these factors may be involved in establishing an appropriate tissue or biochemical environment for the action of *Sry*, given their knockout phenotypes. The sex reversal seen when *M. m. musculus Sry* alleles are present on an *M. m. domesticus* background implicates aberrant interaction between *musculus Sry* and *domesticus* autosomal/X-linked genes [73]. These genes could encode regulators of *Sry* transcription [74, 169, 170] or SRY-interacting proteins [45, 47]. Some of these loci have been mapped [171], and await positional cloning.

Does SRY directly regulate Sox9?

Clearly, *Sox9* is an important gene in the male sex determination pathway. Mutation of *SOX9* can result in complete, male-to-female phenotypic sex reversal in humans. It appears that *Sox9* is intimately linked with Sertoli cell differentiation, even in the absence of *Sry* [150]. The timing and male-specific upregulation of *Sox9* expression in mice are compatible with the possibility that SRY directly regulates *Sox9*. Indeed, a potential binding site for SRY is found in the promoter region of mouse and human *Sox9*, in a 113-bp interval found to confer at least some sex specificity of *Sox9* expression upon reporter constructs transfected into primary mouse fetal testicular and ovarian cells [172]. Even if *Sry* were normally involved in *Sox9* regulation in mammals, evidence exists that it is not absolutely required. In mice, genital ridges of either sex can undergo testicular differentiation when grafted to extragonadal sites such as the kidneys, and testicular tissue induced in this manner expresses *Sox9* [150]. Either these extragonadal sites supply a factor that mimics the action of SRY in regulating *Sox9*, or SRY itself does not regulate *Sox9* expression directly.

The role of *Sox9* in sex determination is not limited to mammals. Male-specific expression of *Sox9* has been reported in chicken and turtle genital ridges, coincident with testis determination in those species [150, 161, 173, 174]. This suggests that *Sox9* is a fundamental component of the male sex-determining pathway in all vertebrates. *Sry* can obviously not be involved in the male-specific expression of *Sox9* in the gonads of non-mammalian species, nor indeed the few mammalian species which lack *Sry*. In classes such as birds, reptiles and fish, different sex-determining switches operate, as described in other papers in this volume. It is generally postulated that sex determination in all vertebrates converges to a common genetic pathway, and that this pathway is activated by different switch mechanisms that can be genetic or temperature-dependent. *Sox9* may represent the point at which these pathways converge. If this is true, then it is possible that this gene is activated by different mechanisms in different vertebrate classes. Alternatively, the pathways may converge at some point upstream from *Sox9*, and similar regulatory mechanisms for *Sox9* might be found in all vertebrates.

SRY, DAX1 and the repression of Sox9

The expression profiles of *Sox9* in male and female gonads suggests that repression of *Sox9* is critical for ovarian development. It is possible to speculate that a repressor of *Sox9* may be expressed in female genital ridges, but that this repressor is itself repressed in male genital ridges, perhaps by *Sry*. In this light it is interesting to note that disruption of a region some 1.4 Mbp upstream of mouse *Sox9* by a transgene results in continued expression (that is, loss of the normal down-regulation) of *Sox9* expression in XX fetal gonads, leading to female-to-male sex reversal (C. Bishop, R. Behringer and P. Overbeek, unpublished data), suggestive of a negative regulatory element in this interval. As noted above, transgenic mice overexpressing *Dax1* have been shown to undergo male-to-female sex reversal [116]. This situation is analogous to duplications of the region including *DAX1*, resulting in male-to-female sex reversal in humans [112], although in this latter case the involvement of alternative or additional genes has not been ruled out. Unlike the situation in humans, however, sex reversal in mice was seen only in combination with a "weak" allele of *Sry* such as Y^{POS}, suggesting that dosage sensitivities differ between the two species. These transgenic experiments suggest that *Dax1* antagonizes *Sry* action [116]. One might speculate that *Dax1* is involved in the suppression of *Sox9*, and that *Sry*'s role is to antagonize *Dax1*. This scenario is really a more refined version of the "*Sry* represses a repressor of maleness" hypothesis of McElreavey and colleagues [35], and it will be of interest to clarify the molecular basis of the antagonism between SRY and DAX1.

Regulation of Sox9

While many alternative models are possible, further clarification of the connection between *Sry* and *Sox9* awaits characterization of the *Sox9* regulatory sequences. As mentioned above, we have made some progress towards this end in mice, implicating sequences in the proximal promoter region as contributing, at least in part, to the sex specificity of *Sox9* expression [172]. It appears that other important elements are scattered for some distance upstream of *SOX9*. CD can be caused by translocations involving breakpoints spread through more than 800 kb upstream from *SOX9* [163, 175]. Wunderle and colleagues [176] have delineated some of the regulatory regions using human *SOX9* YAC-*LacZ* reporter constructs in transgenic mice. While their results indicate that elements required for *SOX9* expression in a variety of tissues are spread through some 350 kb upstream of the gene, they were unable to identify any fragment which directed *LacZ* expression to the gonads. The positive regulatory elements required for expression of *SOX9* in fetal gonads either are not located in the YAC fragments tested, or do not function in mouse cells. If the latter is true, this may once again point to fundamental biochemical differences in the sex-determining pathway between mice and humans.

Beyond Sox9

What lies downstream from *Sox9* in the testis-determining pathway? Compelling evidence from several laboratories, involving a number of *in vitro* and *in vivo* approaches, indicates that SOX9 acts in concert with SF1 to activate *Amh* transcription [102–104, 156]. However, several observations complicate this simple view that the role of *Sox9* in sex determination is to regulate *Amh*. Expression of *Amh* has been found to precede rather than follow that of *Sox9* in chickens and alligators, which would appear to rule out the possibility of *Amh* regulation by SOX9 in those species [177, 178]. Further, other regulators of *Amh* have been identified in recent studies. It has been reported that SF1 associates with two of the four alternately spliced isoforms of WT1, and that this association synergizes the up-regulation of *Amh* [179]. WT1 missense mutations associated with male pseudohermaphroditism fail to synergize with SF1 [179]. The transcription factor GATA4 has also been identified as a regulator of *Amh* expression *in vitro* [126]. These regulatory scenarios do not necessarily negate the role of SOX9 in *Amh* regulation. However, it is clear that SOX9 must have other regulatory targets since *SOX9* mutation results in sex reversal [139, 140] whereas *Amh* mutation does not [97].

Perspectives

Sex determination is fundamental to the survival of any sexually reproducing species. Altering any one of the genetic steps leading to maleness or femaleness would seem to call for simultaneously altering the interdependent steps if that change is to be compatible with normal sexual development, reproduction and hence survival. With this in mind it is indeed astonishing that so many different mechanisms of determining sex can be found among metazoans. The few model organisms studied in any detail provide what is probably only a glimpse of this variability. For example, the pathway used in *Drosophila melanogaster* uses an almost completely different set of genes to those used in *Caenorhabditis elegans* [180]. With the possible exception of *Dmrt1*, it appears that none of the genes used in either of these species is deployed in vertebrate sex determination. This is in stark contrast to many other genetic pathways, such as the *Hox, Wingless/Wnt* and *hedgehog* pathways, that have been structurally and functionally conserved throughout the metazoan subkingdom and perhaps even beyond.

   In trying to draw together data relating to the biochemical and genetic roles of *Sry* and *Sox9* in sex determination, we have seen several examples of differences between vertebrate classes, mammalian genera, rodent species and in some cases even mouse strains, to a degree that far exceeds all other organogenetic systems. To recount just a few examples, there are differences in dosage sensitivities and gene expression patterns, renegade rodents that lack *Sry* altogether, and a panoply of different structural variations of *Sry* among mice. These differences demand that we resist the temptation to put forward unifying models of the genetic control of sex determination. A corollary is that any individual organism will be of only limited value as a "model" of vertebrate or mammalian sex determination. Further study of a number of experimental organisms will undoubtedly reveal further richness of variety.

While sex determination in vertebrates is often described as a pathway, perhaps by analogy with the linear sequence of events in *Drosophila* and *C. elegans*, it is clear that this concept is hopelessly inadequate. Several distinct cell lineages need to be coordinately directed by a series of signalling molecules, receptors, signal transduction systems and transcription factors. A number of factors, such as WT1 and SF1, are clearly used several times in different roles, even within the developing gonad. Recent work has begun to uncover the combinatorial transcription factor interactions involved in various aspects of gonadal development. Clearly, we are dealing with a complex network involving multiple positive and negative regulatory elements. *Sry* and *Sox9* play pivotal roles within this network, but information relating to their modes of action remains sketchy. To date, the majority of efforts to understand the interplay of genes and gene products in sex determination have revolved around attempts at pairwise marriages – SRY and *Amh*, SOX9 and SF1, SOX9 and *Amh*, GATA4 and *Amh*, WT1 and DAX1, and so on. Some of these relationships are no doubt genuine. However, as described above and elsewhere in this volume, some of these studies have resulted in alliances that are likely to be artefactual; others are contradictory or mutually difficult to reconcile. Part of the reason for this is the molecular biologist's reliance on *in vitro* systems, which, while useful, may not always reflect the *in vivo* situation. Another factor is our natural tendency to fit together the pieces of the jigsaw that we have, without regard for the missing pieces. Since we have little more than a handful of transcription factors, it is clear that many more molecules of different classes need to be found. This, and the interconnection of these pieces, will provide the challenge for future research.

Acknowledgements
I am grateful to Gerd Scherer, Andrew Sinclair, Robin Lovell-Badge, Patrick Tam and Vince Harley for helpful comments on an earlier version of the manuscript, and to Jo Bowles, Richard Behringer and Vince Harley for permission to cite unpublished data. This work was supported by grants from the Australian Research Council and the National Health and Medical Research Council of Australia. P. K. is an Australian Research Council Senior Research Fellow.

References

1   Ford CE, Jones KW, Polani PE, de Almeida JC and Briggs JH (1959) A sex chromosome anomaly in a case of gonadal dysgenesis (Turner's Syndrome). *Lancet* i: 711–713

2   Welshons WJ and Russell LB (1959) The Y chromosome as the bearer of male determining factors in the mouse. *Proc Natl Acad Sci USA* 45: 560–566

3   Jacobs PA and Strong JA (1959) A case of human intersexuality having a possible XXY sex determining mechanism. *Nature* 183: 302–303

4   Freguson-Smith MA (1966) X-Y interchange in the aetiology of true hermaphroditism and of XX Klinefelter's syndrome. *Lancet* ii: 475–476

5   Ferguson-Smith MA and Affara NA (1988) Accidental X-Y recombination and the aetiology of XX males and true hermaphrodites. Philos Trans R Soc Lond Biol 322: 133–144

6   Sinclair AH, Berta P, Palmer MS, Hawkins JR, Griffiths BL, Smith MJ et al (1990) A gene from the human sex-determining region encodes a protein with homology to a conserved DNA-binding motif. *Nature* 346: 240–244

7   Gubbay J, Collignon J, Koopman P, Capel B, Economou A, Münsterberg A et al (1990) A gene mapping to the sex-determining region of the mouse Y chromosome is a member of a novel family of embryonically expressed genes. *Nature* 346: 245–250

8   Berta P, Hawkins JR, Sinclair AH, Taylor A, Griffiths BL, Goodfellow PN et al (1990) Genetic evidence equating *SRY* and the male sex determining gene. *Nature* 348: 448–450

9   Jäger RJ, Anvret M, Hall K and Scherer G (1990) A human XY female with a frame shift mutation in the candidate testis-determining gene *SRY*. *Nature* 348: 452–454

10 Koopman P, Gubbay J, Vivian N, Goodfellow P and Lovell-Badge R (1991) Male development of chromosomally female mice transgenic for *Sry. Nature* 351: 117–121

11 Jantzen H-M, Admon A, Bell SP and Tjian R (1990) Nucleolar transcription factor hUBF contains a DNA-binding motif with homology to HMG proteins. *Nature* 344: 830–836

12 van de Wetering M, Oosterwegel M, Dooijes D and Clevers H (1991) Identification and cloning of TCF-1, a T lymphocyte-specific transcription factor containing a sequence-specific HMG box. *EMBO J* 10: 123–132

13 Oosterwegel M, van de Wetering M, Dooijes D, Klomp L, Winoto A, Georgopoulos K et al (1991) Cloning of murine TCF-1, a T cell-specific transcription factor interacting with functional motifs in the CD3-epsilon and T cell receptor alpha enhancers. *J Exp Med* 173: 1133–1142

14 Travis A, Amsterdam A, Belanger C and Grosschedl R (1991) LEF-1, a gene encoding a lymphoid-specific protein with an HMG domain, regulates T-cell receptor α enhancer function. *Genes Dev* 5: 880–894

15 Waterman ML, Fischer WH and Jones KA (1991) A thymus-specific member of the HMG protein family regulates the human T cell receptor Cα enhancer. *Genes Dev* 5: 656–669

16 Harley VR, Jackson DI, Hextall PJ, Hawkins JR, Berkovitz GD, Sockanathan S et al (1992) DNA binding activity of recombinant SRY from normal males and XY females. *Science* 255: 453–456

17 van de Wetering M and Clevers H (1992) Sequence-specific interaction of the HMG box proteins TCF-1 and SRY occurs within the minor groove of a Watson-Crick double helix. *EMBO J* 11: 3039–3044

18 Harley VR, Lovell-Badge R and Goodfellow PN (1994) Definition of a concensus DNA binding site for SRY. *Nucleic Acids Res* 22: 1500–1501

19 Ferrari S, Harley VR, Pontiggia A, Goodfellow PN and Lovell-Badge R (1992) SRY, like HMG1, recognizes sharp angles in DNA. *EMBO J* 11: 4497–4506

20 Giese K, Cox J and Grosschedl R (1992) The HMG domain of lymphoid enhancer factor 1 bends DNA and facilitates assembly of functional nucleoprotein structures. *Cell* 69: 185–195

21 Giese K, Pagel J and Grosschedl R (1994) Distinct DNA-binding properties of the high mobility group domain of murine and human SRY sex-determining factors. *Proc Natl Acad Sci USA* 91: 3368–3374

22 Grossedl R, Giese J and Pagel J (1994) HMG domain proteins: architectural elements in the assembly of nucleoprotein structures. *Trends Genet* 10: 94–100

23 Wolffe AP (1994) Architectural transcription factors. *Science* 264: 1100–1101

24 Giese K, Kingsley C, Kirshner JR and Grosschedl R (1995) Assembly and function of a TCRα enhancer complex is dependent on LEF-1-induced DNA bending and multiple protein-protein interactions. *Genes Dev* 9: 995–1008

25 Pontiggia A, Rimini R, Harley VR, Goodfellow PN, Lovell-Badge R and Bianchi ME (1994) Sex-reversing mutations affect the architecture of SRY-DNA complexes. *EMBO J* 13: 6115–6124

26 Whitfield LS, Lovell-Badge R and Goodfellow PN (1993) Rapid sequence evolution of the mammalian sex determining gene *SRY. Nature* 364: 713–715

27 Pamilo P and O'Neill RJW (1997) Evolution of the Sry genes. *Mol Biol Evol* 14: 49–55

28 Poulat F, Girard F, Chevron M-P, Goze C, Rebillard X, Calas B, Lamb N and Berta P (1995) Nuclear localization of the testis determining gene product SRY. *J Cell Biol* 128: 737–748

29 Südbeck P and Scherer G (1997) Two independent nuclear localization signals are present in the DNA-binding high mobility group domains of *SRY* and *SOX9. J Biol Chem* 272: 27848–27852

30 Desclozeaux M, Poulat F, Barbara PD, Capony JP, Turowski P, Jay P, Mejean C, Moniot B, Boizet B and Berta P (1998) Phosphorylation of an N-Terminal motif enhances DNA-binding activity of the human SRY protein. *J Biol Chem* 273: 7988–7995

31 de la Chapelle A (1981) The etiology of maleness in XX men. *Hum Genet* 58: 105–116

32 Numabe H, Nagafuchi S, Nagahori Y, Tamura T, Kiuchi H, Namiki M et al (1992) DNA analyses of XX and XX-hypospadiac males. *Human Genet* 90: 211–214

33 Fechner PY, Marcantonio SM, Jaswaney V, Stetten G, Goodfellow PN, Migeon CJ et al (1993) The role of the sex determining region Y gene in the etiology of 46,XX maleness. *J Clin Endocrinol Metab* 76: 690–695

34 Meyers-Wallen VN, Schlafer D, Barr I, Lovell-Badge R and Keyzner A (1999) *Sry*-negative XX sex reversal in purebred dogs. *Mol Reprod Dev* 53: 266–273

35 McElreavey K, Vilain E, Herskowitz I and Fellous M (1993) A regulatory cascade hypothesis for mammalian sex determination: SRY represses a negative regulator of male development. *Proc Natl Acad Sci USA* 90: 3368–3372

36 Ng L-J, Wheatley S, Muscat GEO, Conway-Campbell J, Bowles J, Wright E et al (1997) SOX9 binds DNA, activates transcription and co-expresses with type II collagen during chondrogenesis in the mouse. *Dev Biol* 183: 108–121

37 Kamachi Y, Cheah KS and Kondoh H (1999) The mechanism of regulatory target selection by the SOX HMG domain proteins as revealed by comparison of SOX1/2 and SOX9. *Mol Cell Biol* 19: 107–120

38 Poulat F, de Santa Barbara P, Desclozeaux M, Soullier S, Moniot B, Bonneaud N et al (1997) The human testis determining factor *SRY* binds a nuclear factor containing PDZ protein interaction domains. *J Biol Chem* 272: 7167–7172

39 Tajima T, Nakae J, Shinohara N and Fujieda K (1994) A novel mutation localized in the 3′ non-HMG box region of the SRY gene in 46,XY gonadal dysgenesis. *Human Mol Genet* 3: 1187–1189

40 Harley VR, Lovell-Badge R, Goodfellow PN and Hextall PJ (1996) The HMG box of SRY is a calmodulin binding domain. *FEBS Lett* 391: 24–28

41 Gubbay J, Vivian N, Economou A, Jackson D, Goodfellow P and Lovell-Badge R (1992) Inverted repeat structure of the Sry locus in mice. *Proc Natl Acad Sci USA* 89: 7953–7957

42 Dubin RA and Ostrer H (1994) Sry is a transcriptional activator. *Mol Endocrinol* 8: 1182–1192

43 Coward P, Nagai K, Chen D, Thomas HD, Nagamine CM and Lau Y-FC (1994) Polymorphism of a CAG repeat within *Sry* correlates with B6.Y^Dom sex reversal. *Nature Genet* 6: 245–250

44 Dubin RA, Coward P, Lau Y-FC and Ostrer H (1995) Functional comparison of the *Mus musculus molossinus* and *Mus musculus domesticus Sry* genes. *Mol Endocrinol* 9: 1645–1654

45 Bowles J, Berkman J, Cooper L and Koopman P (1999) *Sry* requires a CAG repeat domain for male sex determination in *Mus musculus. Nature Genet* 22: 405–408

46 Lau YFC and Zhang J (1998) Sry interactive protein – implication for the mechanisms of sex determination. *Cytogenet Cell Genet* 80: 128–132

47 Zhang J, Coward P, Xian M and Lau YF (1999) *In vitro* binding and expression studies demonstrate a role for the mouse Sry Q-rich domain in sex determination. *Int J Dev Biol* 43: 219–227

48 Koopman P, Münsterberg A, Capel B, Vivian N and Lovell-Badge R (1990) Expression of a candidate sex-determining gene during mouse testis differentiation. *Nature* 348: 450–452

49 Hacker A, Capel B, Goodfellow P and Lovell-Badge R (1995) Expression of *Sry*, the mouse sex determining gene. *Development* 121: 1603–1614

50 Jeske YWA, Mishina Y, Cohen DR, Behringer RR and Koopman P (1996) Analysis of the role of Amh and Fra1 in the Sry regulatory pathway. *Mol Reprod Dev* 44: 153–158

51 Jeske YWA, Bowles J, Greenfield A and Koopman P (1995) Expression of a linear *Sry* transcript in the mouse genital ridge. *Nature Genet* 10: 480–482

52 Zwingman T, Erickson RP, Boyer T and Ao A (1993) Transcription of the sex-determining region genes *Sry* and *Zfy* in the mouse preimplantation embryo. *Proc Natl Acad Sci USA* 90: 814–817

53 Gutierrez-Adan A, Behboodi E, Murray JD and Anderson GB (1997) Early transcription of the SRY gene by bovine preimplantation embryos. *Mol Reprod Dev* 48: 246–250

54 Fiddler M, Abdel-Rahman B, Rappolee DA and Pergament E (1995) Expression of SRY transcripts in preimplantation human embryos. *Am J Med Genet* 55: 80–84

55 Harry JL, Koopman P, Brennan FE, Graves JAM and Renfree MB (1995) Widespread expression of the testis-determining gene SRY in a marsupial. *Nature Genet* 11: 347–349

56 Payen E, Pailhoux E, Merhi RA, Gianquinto L, Kirszenbaum M, Locatelli A et al (1996) Characterization of ovine SRY transcript and developmental expression of genes involved in sexual differentiation. *Int J Dev Biol* 40: 567–575

57 Clépet C, Schafer AJ, Sinclair AH, Palmer MS, Lovell-Badge R and Goodfellow PN (1993) The human *SRY* transcript. *Human Mol Genet* 2: 2007–2012

58 Capel B, Swain A, Nicolis S, Hacker A, Walter M, Koopman P et al (1993) Circular transcripts of the testis determining gene *Sry* in adult mouse testis. *Cell* 73: 1019–1030

59 Dolci S, Grimaldi P, Geremia R, Pesce M and Rossi P (1997) Identification of a promoter region generating Sry circular transcripts both in germ cells from male adult mice and in male mouse embryonal gonads. *Biol Reprod* 57: 1128–1135

60 Dubin RA, Kazmi MA and Ostrer H (1995) Inverted repeats are necessary for circularization of the mouse testis *Sry* transcript. *Gene* 167: 245–248

61 McLaren A (1988) Somatic and germ-cell sex in mammals. *Philos Trans R Soc Lond Biol* 322: 3–9

62 Merchant H (1975) Rat gonadal and ovarian organogenesis with and without germ cells. An ultrastructural study. *Dev Biol* 44: 1–21

63 McLaren A (1985) Relation of germ cell sex to gonadal development. In: The Origin and Evolution of Sex, pp. 289–300, Halvorson HO and Monroy A (eds), Liss, New York

64 Burgoyne PS, Buehr M, Koopman P, Rossant J and McLaren A (1988) Cell-autonomous action of the testis-determining gene: Sertoli cells are exclusively XY in XX↔XY chimaeric mouse testes. *Development* 102: 443–450

65 Palmer SJ and Burgoyne PS (1991) *In situ* analysis of fetal, prepuberal and adult XX↔XY chimaeric mouse testes: Sertoli cells are predominantly, but not exclusively, XY. *Development* 112: 265–268

66 Rossi P, Dolci S, Albanesi C, Grimaldi P and Geremia R (1993) Direct evidence that the mouse sex-determining gene *Sry* is expressed in the somatic cells of male fetal gonads and in the germ cell line in the adult testis. *Mol Reprod Dev* 34: 369–373

67 Buehr M, Gu S, McLaren A (1993) Mesonephric contribution to testis differentiation in the fetal mouse. *Development* 117: 273–281

68 Martineau J, Nordqvist K, Tilmann C, Lovell-Badge R, Capel B (1997) Male-specific cell migration into the developing gonad. *Curr Biol* 7: 958–968

69 Capel B, Albrecht KH, Washburn LL, Eicher EM (1999) Migration of mesonephric cells into the mammalian gonad depends on *Sry*. *Mech Dev* 84: 127–131

70 Tilmann C, Capel B (1999) Mesonephric cell migration induces testis cord formation and Sertoli cell differentiation in the mammalian gonad. *Development* 126: 2883–2890

71 Palmer MS, Sinclair AH, Berta P, Ellis NA, Goodfellow PN, Abbas NE et al (1989) Genetic evidence that ZFY is not the testis-determining factor. *Nature* 342: 937–939

72 Capel B, Rasberry C, Dyson J, Bishop CE, Simpson E, Vivian N et al (1993) Deletion of Y chromosome sequences located outside the testis determining region can cause XY female sex reversal. *Nature Genet* 5: 301–307

73 Eicher EM, Washburn LL, Whitney JB and Morrow KE (1982) *Mus poschiavinus* Y chromosome in the *C57BL/6J* murine genome causes sex reversal. *Science* 217: 535–537

74 Nagamine CM, Morohashi K-I, Carlisle C and Chang DK (1999) Sex reversal caused by *Mus musculus domesticus* Y chromosomes linked to variant expression of the testis-determining gene *Sry. Dev Biol* 216: 182–194

75 Veitia RA, Fellous M and McElreavey K (1997) Conservation of Y chromosome-specific sequences immediately 5′ to the testis determining gene in primates. *Gene* 199: 63–70

76 Margarit E, Guillen A, Rebordosa C, Vidal-Taboada J, Sanchez M, Ballesta F et al (1998) Identification of conserved potentially regulatory sequences of the Sry gene from 10 different species of mammals. *Biochem Biophys Res Commun* 245: 370–377

77 Schmitt-Ney M, Thiele H, Kaltwasser P, Bardoni B, Cisternino M and Scherer G (1995) Two novel SRY missense mutations reducing DNA binding identified in XY females and their mosaic fathers. *Am J Hum Genet* 56: 862–869

78 Kwok C, Tyler-Smith C, Mendonca BB, Hughes I, Berkovitz GD, Goodfellow PN et al (1996) Mutation analysis of the 2 kb 5′ to SRY in XY females and XY intersex subjects. *J Med Genet* 33: 465–468

79 Veitia R, Ion A, Barbaux S, Jobling MA, Souleyreau N, Ennis K et al (1997) Mutations and sequence variants in the testis-determining region of the Y chromosome in individuals with a 46,XY female phenotype. *Hum Genet* 99: 648–652

80 Poulat F, Desclozeaux M, Tuffery S, Jay P, Boizet B and Berta P (1998) Mutation in the 5′ noncoding region of the SRY gene in an XY sex-reversed patient. *Hum Mut* S192–S194

81 McElreavey K, Vilain E, Abbas N, Costa JM, Souleyreau N, Kucheria K et al (1992) XY sex reversal associated with a deletion 5′ to the *SRY* "HMG box" in the testis-determining region. *Proc Natl Acad Sci USA* 89: 11016–11020

82 McElreavey K, Vilain E, Barbaux S, Fuqua JS, Fechner PY, Souleyreau N et al (1996) Loss of sequences 3′ to the testis-determining gene, *SRY*, including the Y pseudoautosomal boundary associated with partial testicular determination. *Proc Natl Acad Sci USA* 93: 8590–8594

83 Griffiths R (1991) The isolation of conserved DNA sequences related to the human sex-determining region Y gene from the lesser black-backed gull (*Larus fuscus*). *Philos Trans R Soc Lond Biol* 244: 123–128

84 Tiersch TR, Mitchell MJ and Wachtel SS (1991) Studies on the phylogenetic conservation of the SRY gene. *Hum Genet* 87: 571–573

85 Just W, Rau W, Vogul W, Akhverdian M, Fredga K, Graves JAM et al (1995) Absence of *Sry* in species of the vole *Ellobius. Nature Genet* 11: 117–118

86 Soullier S, Hanni C, Catzeflis F, Berta P and Laudet V (1998) Male sex determination in the spiny rat *Tokudaia osimensis* (Rodentia: Muridae) is not Sry dependent. *Mamm Genome* 9: 590–592

87 Stevanovic M, Lovell-Badge R, Collignon J and Goodfellow PN (1993) *SOX3* is an X-linked gene related to *SRY. Human Mol Genet* 2: 2013–2018

88 Foster JW and Graves JAM (1994) An SRY-related sequence on the marsupial X chromosome: implications for the evolution of the mammalian testis determining gene. *Proc Natl Acad Sci USA* 91: 1927–1931

89 Collignon J, Sockanathan S, Hacker A, Cohen-Tannoudji M, Norris D, Rastan S et al (1996) A comparison of the properties of *Sox-3* with *Sry* and two related genes, *Sox-1* and *Sox-2. Development* 122: 509–520

90 Uwanogho D, Rex M, Cartwright EJ, Pearl G, Healy C, Scotting P et al (1995) Embryonic expression of the chicken Sox2, Sox3 and Sox11 genes suggests an interactive role in neural development. *Mech Dev* 49: 23–36

91 Penzel R, Oschwald R, Chen YL, Tacke L and Grunz H (1997) Characterization and early embryonic expression of a neural specific transcription factor xSox2 in *Xenopus laevis. Int J Dev Biol* 41: 667–677

92 McBride D, Sang H and Clinton M (1997) Expression of *Sry*-related genes in the developing genital ridge/mesonephros of the chick embryo. *J Reprod Fertil* 109: 59–63

93 Smith CA, Smith MJ and Sinclair AH (1999) Gene expression during gonadogenesis in the chicken embryo. *Gene* 234: 395–402

94 Graves JA (1998) Interactions between SRY and SOX genes in mammalian sex determination. *BioEssays* 20: 264–269

95   Mumm S, Zucchi I and Pilia G (1997) SOX3 gene maps ear DXS984 in Xq27.1, within
     candidate regions for several X-linked disorders. *Am J Med Genet* 72: 376–378
96   Swain A, Lovell-Badge R (1999) Mammalian sex determination: a molecular drama.
     Genes Dev 13: 755–767
97   Behringer RR, Finegold MJ and Cate RL (1994) Müllerian-Inhibiting Substance function
     during mammalian sexual development. *Cell* 79: 415–425
98   Ikeda Y, Shen WH, Ingraham HA and Parker KL (1994) Developmental expression
     of mouse steroidogenic factor-1, an essential regulator of the steroid hydroxylases. *Mol
     Endocrinol* 8: 654–662
99   Hatano O, Takayama K, Imai T, Waterman MR, Takakusu A, Omura T et al (1994) Sex-
     dependent expression of a transcription factor, Ad4BP, regulating steroidogenic P-450
     genes in the gonads during prenatal and postnatal rat development. *Development* 120:
     2878–2797
100 Luo X. Ikeda Y and Parker KL (1994) A cell-specific nuclear receptor is essential for
     adrenal and gonadal development and sexual differentiation. *Cell* 481–490
101 Ackermann JC, Ito M, Ito M, Hindmarsh PC and Jameson JL (1999) A mutation in the
     gene encoding steroidogenic factor-1 causes XY sex reversal and adrenal failure in humans.
     Nature Genet 22: 125–126
102 Shen W-H, Moore CCD, Ikeda Y, Parker KL and Ingraham H (1994) Nuclear receptor
     steriodogenic factor 1 regulates the Müllerian inhibiting substance gene: a link to the sex
     determination cascade. *Cell* 77: 651–661
103 Giuili G, Shen W-H and Ingraham HA (1997) The nuclear receptor SF-1 mediates sexually
     dimorphic expression of müllerian inhibiting substance, *in vivo. Development* 124: 1799–
     1807
104 Arango NA, Lovell-Badge R and Behringer RR (1999) Targeted mutagenesis of the endo-
     genous mouse *Mis* gene promoter: *In vivo* definition of genetic pathways of vertebrate
     sexual development. *Cell* 99: 409–419
105 Kreidberg JA, Sariola H, Loring JM, Maeda M, Pelletier J, Housman D et al (1993) WT-1
     is required for early kidney development. *Cell* 74: 679–691
106 Pelletier J, Bruening W, Kashtan CE, Mauer SM, Manivel JC, Striegel JE et al (1991)
     Germline mutations in the Wilms' tumor suppressor gene are associated with abnormal
     urogenital development in Denys-Drash syndrome. *Cell* 67: 437–447
107 Behlke MA, Bogan JS, Beer-Romero P and Page DC (1993) Evidence that the SRY
     protein is encoded by a single exon on the human Y chromosome. *Genomics* 17: 736–739
108 Parker KL, Schedl A and Schimmer BP (1999) Gene interactions in gonadal development.
     Ann Rev Physiol 61: 417–433
109 Kim J, Prawitt D, Bardeesy N, Torban E, Vicaner C, Goodyer P, Zabel B and Pelletier J
     (1999) The Wilms' tumor suppressor gene (WT1) product regulates Dax-1 gene expression
     during gonadal differentiation. *Mol Cell Biol* 19: 2289–2299
110 Miyamoto N, Yoshida M, Kuratani S, Matsuo I, Aizawa S (1997) Defects of urogenital
     development in mice lacking *Emx2. Development* 124: 1653–1664
111 Shawlot W and Behringer RR (1995) Requirement for *Lim1* in head-organizer function.
     Nature 374: 425–430
112 Bardoni B, Zanaria E, Guioli S, Floridia G, Worley KC, Tonini G et al (1994) A dosage
     sensitive locus at chromosome Xp21 is involved in male to female sex reversal. *Nature
     Genet* 7: 497–501
113 Muscatelli F, Strom TM, Walker AP, Zanaria E, Recan D, Meindl A et al (1994) Mutations
     in the *DAX-1* gene give rise to both X-linked adrenal hypoplasia congenita and hypogona-
     dotropic hypogonadism. *Nature* 372: 672–676
114 Zanaria E, Muscatelli F, Bardoni B, Strom TM, Guioli S, Guo W et al (1994) An unusual
     member of the nuclear hormone receptor superfamily responsible for X-linked adrenal
     hypoplasia congenita. *Nature* 372: 635–641
115 Swain A, Zanaria E, Hacker A, Lovell-Badge R and Camerino G (1996) Mouse *Dax1*
     expression is consistent with a role in sex determination as well as adrenal and hypothala-
     mus function. *Nature Genet* 12: 404–409
116 Swain A, Narvaez V, Burgoyne PS, Camerino G and Lovell-Badge R (1998) *Dax1* ant-
     agonizes *Sry* action in mammalian sex determination. *Nature* 391: 761–767
117 Yu RN, Ito M, Saunders TL, Camper SA and Jameson JL (1998) Role of *Ahch* in gonadal
     development and gametogenesis. *Nature Genet* 20: 353–357

118 Parker KL and Schimmer BP (1998) *Ahch* and the feminine mystique. *Nature Genet* 20: 318–319

119 Lalli E, Bardoni B, Zazopoulos E, Wurtz J-M, Strom TM, Moras D and Sassone-Corsi P (1997) A transcriptional silencing domain in DAX-1 whose mutation causes adrenal hypoplasia congenita. *Mol Endocrinol* 11: 1950–1960

120 Ito M, Yu R and Jameson JL (1997) DAX-1 inhibits SF-1-mediated transactivation via a carboxy-terminal domain that is deleted in adrenal hypoplasia congenita. *Mol Cell Biol* 17: 1476–1483

121 Zazopoulos E, Lalli E, Stocco DM, Sassone-Corsi P (1997) DNA binding and transcriptional repression by DAX-1 blocks steroidogenesis. *Nature* 390: 311–315

122 Sandhoff TW and McLean MP (1999) Repression of the rat steroidogenic acute regulatory (StAR) protein gene by PGF2alpha is modulated by the negative transcription factor DAX- 1. *Endocrine* 10: 83–91

123 Haqq CM, King C-Y, Donahoe PK and Weiss MA (1993) SRY recognizes conserved DNA sites in sex-specific promoters. *Proc Natl Acad Sci USA* 90: 1097–1101

124 Cohen DR, Sinclair AH and McGovern JD (1994) Sry protein enhances transcription of Fos-related antigen 1 promoter constructs. *Proc Natl Acad Sci USA* 91: 4372–4376

125 Katoh-Fukui Y, Tsuchiya R, Shiroishi T, Nakahara Y, Hashimoto N, Noguchi K and Higashinakagawa T (1998) Male-to-female sex reversal in M33 mutant mice. *Nature* 393: 688–692

126 Viger RS, Mertineit C, Trasler JM and Nemer M (1998) Transcription factor GATA-4 is expressed in a sexually dimorphic pattern during mouse gonadal development and is a potent activator of the Müllerian inhibiting substance promoter. *Development* 125: 2665–2675

127 Tremblay JJ and Viger RS (1999) Transcription factor GATA-4 enhances Mullerian inhibiting substance gene transcription through a direct interaction with the nuclear receptor SF-1. *Mol Endocrinol* 13: 1388–1401

128 Raymond CS, Shamu CE, Shen MM, Seifert KJ, Hirsch B, Hodgkin J and Zarkower D (1998) Evidence for evolutionary conservation of sex-determining genes. *Nature* 391: 691–695

129 Raymond CS, Parker ED, Kettlewell JR, Brown LG, Page DC, Kusz K, Jaruzelska J, Reinberg Y, Flejter WL, Bardwell VJ et al (1999) A region of human chromosome 9p required for testis development contains two genes related to known sexual regulators. *Human Mol Genet* 8: 989–996

130 Nanda I, Shan Z, Schartl M, Burt DW, Koehler M, Nothwang H, Grutzner F, Paton IR, Windsor D, Dunn I et al (1999) 300 million years of conserved synteny between chicken Z and human chromosome 9. *Nature Genet* 21: 258–259

131 Raymond CS, Kettlewell JR, Hirsch B, Bardwell VJ and Zarkower D (1999) Expression of *Dmrt1* in the genital ridge of mouse and chicken embryos suggests a role in vertebrate sexual development. *Dev Biol* 215: 208–220

132 Smith CA, McClive PJ, Western PS, Reed KJ, Sinclair AH (1999) Conservation of a sex-determining gene. *Nature* 402: 601–602

133 Wegner M (1999) From head to toes: the multiple facets of Sox proteins. *Nucleic Acids Res* 27: 1409–1420

134 Wright EM, Snopek B and Koopman P (1993) Seven new members of the *Sox* gene family expressed during mouse development. *Nucleic Acids Res* 21: 744

135 Pevny LH and Lovell-Badge R (1997) Sox genes find their feet. *Curr Opin Genet Dev* 7: 338–344

136 Schilham MW, Oosterwegel MA, Moerer P, Ya J, de Boer PAJ, van de Wetering M, Verbeek S, Lamers WH, Kruisbeek AM, Cumano A et al (1996) Defects in cardiac outflow tract formation and pro-B-lyphocyte expansion in mice lacking *Sox-4*. *Nature* 380: 711–714

137 Nishiguchi S, Wood H, Kondoh H, Lovell-Badge R and Episkopou V (1998) Sox1 directly regulates the gamma-crystallin genes and is essential for lens development in mice. *Genes Dev* 12: 776–781

138 Bi W, Deng JM, Zhang Z, Behringer RR and de Crombrugghe B (1999) *Sox9* is required for cartilage formation. *Nature Genet* 22: 85–89

139 Foster JW, Dominguez-Steglich MA, Guioli S, Kwok C, Weller PA, Weissenbach J et al (1994) Campomelic dysplasia and autosomal sex reversal caused by mutations in an *SRY*-related gene. *Nature* 372: 525–530

140 Wagner T, Wirth J, Meyer J, Zabel B, Held M, Zimmer J et al (1994) Autosomal sex reversal and campomelic dysplasia are caused by mutations in and around the *SRY*-related gene *SOX9*. *Cell* 79: 1111–1120

141 Pingault V, Bondurand N, Kuhlbrodt K, Goerich DE, PrEhu M-O, Puliti A et al (1998) *SOX10* mutations in patients with Waardenburg-Hirschsprung disease. *Nature Genet* 18: 171–173

142 Wright E, Hargrave MR, Christiansen J, Cooper L, Kun J, Evans T et al (1995) The *Sry*-related gene *Sox-9* is expressed during chondrogenesis in mouse embryos. *Nature Genet* 9: 15–20

143 Houston CS, Opitz JM, Spranger JW, Macpherson RI, Reed MH, Gilbert EF et al (1983) The campomelic syndrome: review, report of 17 cases and follow-up on the currently 17-year old boy first reported by Maroteaux et al in 1971. *Am J Med Genet* 15: 3–28

144 Tommerup N, Schempp W, Mienecke P, Pedersen S, Bolund L, Brandt C et al (1993) Assignment of an autosomal sex reversal locus (SRA1) and campomelic dysplasia (CMPD1) to 17q24.3-q25.1. *Nature Genet* 4: 170–174

145 Bell DM, Leung KKH, Wheatley SC, Ng LJ, Zhou S, Ling KW et al (1997) SOX9 directly regulates the type-II collagene. *Nature Genet* 16: 174–178

146 Lefebvre V, Huang W, Harley VR, Goodfellow PN and De Crombrugghe B (1997) SOX9 is a potent activator of the chondrocyte-specific enhancer of the pro-alpha-1(II) collagen gene. *Mol Cell Biol* 17: 2336–2346

147 Kwok C, Weller P, Guioli S, Foster JW, Mansour S, Zuffardi O et al (1995) Mutations in *SOX9*, the gene responsible for campomelic dysplasia and autosomal sex reversal. *Am J Hum Genet* 57: 1028–1036

148 Meyer J, Südbeck P, Held M, Wagner T, Schmitz ML, Bricarelli FD et al (1997) Mutational analysis of the SOX9 gene in campomelic dysplasia and autosomal sex reversal: lack of genotype/phenotype correlations. *Human Mol Genet* 6: 91–98

149 Goji K, Nishijima E, Tsugawa C, Nishio H, Pokharel R and Matsuo M (1998) Novel missense mutation in the HMG box of SOX9 gene in a Japanese XY male resulted in campomelic dysplasia and severe defect in masculinization. *Hum Mut Suppl* 1: S114–S116

150 Morais da Silva S, Hacker A, Harley V, Goodfellow P, Swain A and Lovell-Badge R (1996) *Sox9* expression during gonadal development implies a conserved role for the gene in testis differentiation in mammals and birds. *Nature Genet* 14: 62–68

151 Südbeck P, Schmitz ML, Bauerle PA and Scherer G (1996) Sex reversal by loss of the C-terminal transactivation domain of human SOX9. *Nature Genet* 13: 230–232

152 McDowall S, Argentaro A, Ranganathan S, Weller P, Mertin S, Mansour S, Tolmie J and Harley V (1999) Functional and structural studies of wild type SOX9 and mutations causing campomelic dysplasia. *J Biol Chem* 274: 24023–24030

153 Sekiya I, Koopman P, Watanabe H, Ezura Y, Yamada Y and Noda M (1997) SOX9 enhances aggrecan gene expression via the promoter region containing a single HMG box sequence in a chondrogenic cell line, TC6. *J Bone Miner Res* 12: P222

154 Xie WF, Zhang X, Sakano S, Lefebvre V and Sandell LJ (1999) Trans-activation of the mouse cartilage-derived retinoic acid-sensitive protein gene by Sox9. *J Bone Miner Res* 14: 757–763

155 Lefebvre V, Li P and de Crombrugghe B (1998) A new long form of Sox5 (L-Sox5), Sox6 and Sox9 are coexpressed in chondrogenesis and cooperatively activate the type II collagen gene. *EMBO J* 17: 5718–5733

156 de Santa Barbara P, Bonneaud N, Boizet B, Desclozeaux M, Moniot B, Südbeck P, Scherer G, Poulat F and Berta P (1998) Direct interaction of SRY-related protein SOX9 and steroidogenic factor 1 regulates transcription of the human anti-Müllerian hormone gene. *Mol Cell Biol* 18: 6653–6665

157 Schafer AJ, Dominguez-Steglich MA, Guioli S, Kwok C, Weller PA, Stevanovic M et al (1995) The role of SOX9 in autosomal sex reveral and campomelic dysplasia. *Philos Trans R Soc Lond Biol* 350: 271–278

158 Kwok C, Goodfellow PN and Hawkins JR (1996) Evidence to exclude SOX9 as a candidate gene for XY sex reversal without skeletal malformation. *J Med Genet* 33: 800–801

159 Zhao Q, Eberspaecher H, Lefebvre V and De Crombrugghe B (1997) Parallel expression of *Sox9* and *Col2al* in cells undergoing chondrogenesis. *Dev Dyn* 209: 377–386

160 Healy C, Uwanogho D and Sharpe PT (1996) Expression of the chicken *Sox9* gene marks the onset of cartilage differentiation. *Ann NY Acad Sci* 785: 261–262

161 Kent J, Wheatley SC, Andrews JE, Sinclair AH and Koopman P (1996) A male-specific role for SOX9 in vertebrate sex determination. *Development* 122: 2813–2822

162 Y-chromosome activity – a developmental study of X-linked transgene activity in sex-reversed X/XSxr(a) mouse embryos. *Dev Biol* 199: 235–244

163 Wirth J, Wagner T, Meyer J, Pfeiffer RA, Tietze HU, Schempp W et al (1996) Transloca-tion breakpoints in three patients with campomelic dysplasia and autosomal sex reversal map more than 130 kb from SOX9. *Hum Genet* 97: 186–193

164 Huang B, Wang S, Ning Y, Lamb AN and Bartley J (1999) Autosomal XX sex reversal caused by duplication of *SOX9*. *Am J Med Genet* 87: 349–353

165 Goodfellow PN and Lovell-Badge R (1993) *Sry* and sex determination in mammals. *Ann Rev Genet* 27: 71–92

166 McElreavey K and Fellous M (1997) Sex-determining genes. *Trends Endocrinol Metab* 8: 342–346

167 Greenfield A (1998) Genes, cells and organs: recent developments in the molecular genetics of mammalian sex determination. *Mamm Genome* 9: 683–687

168 Capel B (1998) Sex in the 90s – Sry and the switch to the male pathway. *Ann Rev Physiol* 60: 497–523

169 Eicher EM, Shown EP and Washburn LL (1995) Sex reversal in C57BL/6J-Y-POS mice corrected by a Sry transgene. *Philos Trans R Soc Lond Biol* 350: 263–269

170 Albrecht KH and Eicher EM (1997) DNA sequence analysis of Sry alleles (subgenus Mus) implicates misregulation as the cause of C57BL/6J-Y-pos sex reversal and defines the Sry functional unit. *Genetics* 147: 1267–1277

171 Eicher EM, Washburn LL, Schork NJ, Lee BK, Shown EP, Xu X et al (1996) Sex-deter-mining genes on mouse autosomes identified by linkage analysis of C57BL/6J-Y[POS] sex reversal. *Nature Genet* 14: 206–209

172 Kanai Y and Koopman P (1999) Structural and functional characterization of the mouse *Sox9* promoter: implications for campomelic dysplasia. *Human Mol Genet* 8: 691–696

173 Spotila LD, Spotila JR and Hall SE (1998) Sequence and expression analysis of WT1 and Sox9 in the red-eared slider turtle, *Trachemys scripta. J Expt Zool* 281: 417–427

174 Moreno-Mendoza N, Harley VR and Merchant-Larios H (1999) Differential expression of SOX9 in gonads of the sea turtle Lepidochelys olivacea at male- or female-promoting temperatures. *J Expt Zool* 284: 705–710

175 Pfeifer D, Kist R, Dewar K, Devon K, Lander ES, Birren B, Korniszewski L, Back E and Scherer G (1999) Campomelic dysplasia translocation breakpoints are scattered over 1 Mb proximal to SOX9: evidence for an extended control region. *Am J Hum Genet* 65: 111–124

176 Wunderle VM, Critcher R, Hastie N, Goodfellow PN and Schedl A (1998) Deletion of long-range regulatory elements upstream of SOX9 causes campomelic dysplasia. *Proc Natl Acad Sci USA* 95: 10649–10654

177 Oreal E, Pieau C, Mattei MG, Josso N, Picard JY, Carré-Eusebe D et al (1998) Early expression of Amh in chicken embryonic gonads precedes testicular Sox9 expression. *Dev Dyn* 212: 522–532

178 Western PS, Harry JL, Graves JA and Sinclair AH (1999) Temperature-dependent sex determination: upregulation of SOX9 expression after commitment to male development. *Dev Dynamics* 214: 171–177

179 Nachtigal MW, Hirokawa Y, Enyeart-van Houten DL, Flanagan JN, Hammer GD and Ingraham HA (1998) Wilms' tumor 1 and Dax1 modulate the orphan nuclear receptor SF1 in sex-specific gene expression. *Cell* 93: 445–454

180 Hodgkin J (1990) Sex determination compared in *Drosophila* and *Caenorhabditis. Nature* 344: 721–728

181 Werner MH, Huth JR, Gronenborn AM and Clore GM (1995) Molecular basis of human 46 X,Y sex reversal revealed from the three-dimensional solution structure of the human SRY-DNA complex. *Cell* 81: 705–714

182 Jäger R, Harley V, Pfeiffer R, Goodfellow P and Scherer G (1992) A familial mutation in the testis-determining gene SRY shared by both sexes. *Hum Genet* 90: 350–355

183 Hawkins J, Taylor A, Goodfellow P, Migeon C, Smith K and Berkovitz G (1992) Evidence for increased prevalence of SRY mutations in XY females with complete rather than partial gonadal dysgenesis. *Am J Hum Genet* 51: 979–984

184 Muller J, Schwartz M and Skakkebaek N (1992) Analysis of the sex-determining region of the Y chromosome (SRY) in sex reversed patients: point-mutation in SRY causing sex-reversion in a 46,XY female. *J Clin Endocrinol Metab* 75: 331–333

185 McElreavey KD, Vilain E, Boucekkine C, Vidaud M, Jaubert F, Richaud F et al (1992) XY sex reversal associated with a nonsense mutation in SRY. *Genomics* 13: 838–840

186 Zeng Y, Ren Z, Zhang M, Huang Y, Zeng F and Huang S (1993) A new *de novo* mutation (A113T) in HMG box of the SRY gene leads to XY gonadal dysgenesis. *J Med Genet* 30: 655–657

187 Affara NA, Chalmers IJ and Ferguson-Smith MA (1993) Analysis of the SRY gene in 22 sex-reversed XY females identifies four new point mutations in the conserved DNA binding domain. *Human Mol Genet* 2: 785–789

188 Braun A, Kammerer S, Cleve H, Lohrs U, Schwarz H and Kuhnle U (1993) True hermaphroditism in a 46,XY individual, caused by a postzygotic somatic point mutation in the male gonadal sex-determining locus (SRY): molecular genetics and histological findings in a sporadic case. *Am J Hum Genet* 52: 578–585

189 Poulat F, Soullier S, Gozé C, Heitz F, Calas B and Berta P (1994) Description and functional implications of a novel mutation in the sex-determining gene SRY. *Hum Mut* 3: 20–204

190 Iida T, Nakahori Y, Komaki R, Mori E, Hayashi N, Tsutsumi O et al (1994) A novel nonsense mutation in the HMG box of the SRY gene in a patient with XY sex reversal. *Human Mol Genet* 3: 1437–1438

191 Hiort O, Gramss B and Klauber G (1995) True hermaphroditism with 46,XY karyotype and a point mutation in the SRY gene. *J Pediatr* 126: 1022

192 Bilbao JR, Loridan L and Castano L (1996) Novel postzygotic nonsense mutation in SRY in familial XY gonadal dysgenesis. *Hum Genet* 97: 537–539

193 Battiloro E, Angeletti B, Tozzi MC, Bruni L, Tondini S, Vignetti P et al (1997) A novel double nucleotide substitution in the HMG box of the SRY gene associated with Swyer syndrome. *Hum Genet* 100: 585–587

194 Hines RS, Tho SPT, Zhang YY, Plouffe L Jr, Hansen KA, Khan I et al (1997) Paternal somatic and germ-line mosaicism for a sex-determining region on Y (SRY) missense mutation leading to recurrent 46,XY sex reversal. *Fertil Steril* 67: 675–679

195 Cameron FJ, Smith MJ, Warne GL and Sinclair AH (1998) Novel mutation in the Sry gene results in 46,XY gonadal dysgenesis. *Hum Mut Suppl* 1: S110–S111

196 Scherer G, Held M, Erdel M, Meschede D, Horst J, Lesniewicz R et al (1998) Three novel SRY mutations in XY gonadal dysgenesis and the enigma of XY gonadal dysgenesis cases without SRY mutations. *Cytogenet Cell Genet* 80: 188–192

197 Domenice S, Yumie Nishi M, Correia Billerbeck A, Latronico A, Aparecida Medeiros M, Russel A et al (1998) A novel missense mutation (S18N) in the 5′ non-HMG box region of the SRY gene in a patient with partial gonadal dysgenesis and his normal male relatives. *Hum Genet* 102: 213–215

198 Brown S, Yu C, Lanzano P, Heller D, Thomas L, Warburton D et al (1998) A *de novo* mutation (Gln2Stop) at the 5′ end of the SRY gene leads to sex reversal with partial ovarian function. *Am J Hum Genet* 62: 189–192

199 Dörk T, Stuhrmann M, Miller K and Schmidtke J (1998) Independent observation of SRY mutation I90M in a patient with complete gonadal dysgenesis. *Hum Mut* 11: 90–91

200 Takagi A, Imai A and Tamaya T (1999) A novel sex-determining region on Y (*SRY*) nonsense mutation identified in a 45,X/47,XYY female. *Fertil Steril* 72: 167–169

201 Graves PE, Davis D, Erickson RP, Lopez M, Kofman-Alfaro S, Mendez JP and Speer IE (1999) Ascertainment and mutational studies of *SRY* in nine XY females. *Am J Med Genet* 83: 138–139

202 Kwok C, Goodwin LL, Sillence DO et al (1996) A novel germ line mutation in *SOX9* causes familial campomelic dysplasia and sex reversal. *Human Mol Genet* 5: 1625–1630

203 Hageman RM, Cameron FJ and Sinclair AH (1998) Mutation analysis of the *SOX9* gene in a patient with campomelic dysplasia. *Hum Mut Suppl* 1: S112–S113

204 Takamatsu N, Kanda H, Ito M, Yamashita A, Yamashita S and Shiba T (1997) Rainbow trout SOX9: cDNA cloning, gene structure and expression. *Gene* 202: 167–170

Genes and Mechanisms in Vertebrate Sex Determination
ed. by G. Scherer and M. Schmid
© 2001 Birkhäuser Verlag Basel/Switzerland

DAX-1, an "antitestis" gene

Peter N. Goodfellow[1] and Giovanna Camerino[2]

[1] *Department of Genetics, University of Cambridge, Downing Street, Cambridge CB2 3EH, UK*
[2] *Biologia Generale e Genetica Medica, Universitá di Pavia, via Forlanini 14, I-27100 Pavia, Italy*

Summary. The *DAX-1* gene has been involved in the dosage sensitive sex reversal (DSS) phenotype, a male-to-female sex-reversal syndrome due to the duplication of a small region of human chromosome Xp21. *Dax-1* and *Sry* have been shown to act antagonistically in the mouse system, where increasing expression of the former leads to female development and increasing activity of the latter to male development. Although these data strongly implicate *DAX-1* in sex determination, the mouse and human proteins appear to behave differently. Absence of DAX-1 is responsible for adrenal hypoplasia congenita, a human inherited disorder characterized by adrenal insufficiency and hypogonadotropic hypogonadism. Unlike human patients, *Dax-1*-deficient XY mice have normal levels of corticotropins and adrenal hormones but are sterile. *Dax-1*-deficient females are fertile. The DAX-1 protein, an unusual member of the nuclear hormone receptor, may act as a transcriptional repressor. It has been shown to both repress transcriptional activators by direct protein-protein interactions and to bind DNA hairpin structures and repress target genes.

DAX-1 function

DAX-1 and the DSS phenotype

The *DAX-1* gene [1] was isolated from the DSS (dosage sensitive sex reversal) locus, a region of the short arm of the human X chromosome (Xp21) [2]. XY individuals, with a duplication of DSS, exhibit male-to-female sex reversal and gonadal dysgenesis, despite the presence of an intact *SRY* gene. Although XXY individuals develop as males, the single X chromosome in patients with DSS duplications does not undergo X inactivation, implying that the sex-reversal gene is both dosage-sensitive and normally subject to X inactivation. XY patients with deletions of the entire DSS region develop as males [2]. Taken together, these observations suggest that a gene(s) within the DSS region disrupts testis formation when present in a double dose but it is not required for normal testis development. It was proposed that this gene (or genes) is an "antitestis gene" that may be required for ovarian development [2].

The smallest duplication, found in sex-reversed patients, is a 160-kb region of Xp21. This region contains, in addition to the *DAX-1* gene, the *MAGEb/DAM* genes: a family of four to five genes related to the *MAGE* gene family. *MAGE* genes encode tumour-associated antigens of unknown function [3, 4]. As described below, recent biochemical and genetic experiments have equated *DAX-1* with the sex-reversal gene DSS.

Dax-1 overexpression in the mouse

The hypothesis that *DAX-1* is equivalent to DSS was tested by over expression of *Dax-1* (the murine homologue of *DAX-1*) in transgenic mice. Mice were constructed with extra copies of the *Dax-1* gene [5]. High levels of exogenous *Dax-1* expression in the genital ridge retards testis differentiation in XY mice with an *Sry* allele of *Mus musculus musculus* origin, but complete sex reversal occurred when other alleles of *Sry* were used, either the naturally occurring *M. poschiavinus* allele or an *Sry* transgene. This supports the hypothesis that *Dax-1* and *Sry* act antagonistically, where increasing expression of the former leads to female development and increasing activity of the latter to male development.

Although these data strongly implicate *Dax-1* in sex determination, the mouse and human system appear to be behaving differently. A double dose of human *DAX-1* is sufficient to sex-reverse patients with duplications of Xp21, but much higher levels of expression seem to be required for the mouse *Dax-1* gene to compete effectively with an *Sry* allele of *M. m. musculus* origin. One explanation would be the existence of an additional gene(s) in the 160-kb minimal DSS region that cooperates with *DAX-1* to achieve complete sex reversal when present in double dose. A more plausible explanation is that *Sry* and *Dax-1* show species differences in levels of expression relative to critical thresholds. For example, the *M. m. musculus Sry* gene could be expressed significantly higher than the minimum level required to induce testes. Perhaps the human *SRY* gene is similar to the *M. poschiavinus Sry* gene that appears to be expressed at a level much closer to its threshold.

Another possible species difference is in the precise timing of expression of the two genes. If *Dax-1* expression is delayed with respect to *Sry* in *M.m. musculus*, it may never reach the level required to compete before *Sry* has reached its critical threshold and initiated Sertoli cell differentiation. If *Sry* expression is delayed, high levels of *Dax-1* would be able to act first. Testis formation in transgenic XX *Dax:Sry* embryos is retarded in comparison to that in normal XY embryos. This is consistent with a delay in *Sry* expression when it is driven by the *Dax-1* regulatory region and may explain why *Dax-1* competes effectively with *Sry* when the two genes are expressed from identical regulatory sequences.

Finally, subsequent events in gonadal differentiation may differ between humans and mice. In the mouse, fetal ovo-testes usually resolve into mature testes by birth. Perhaps in humans, the bias is towards maintaining ovarian (or streak) development. If this were the case, more examples of sex reversal would be detected in humans than mice.

DAX-1 absence

DAX-1 functional absence in XY individuals. Absence of a functional DAX-1 protein is responsible for the X-linked form of adrenal hypoplasia congenita (AHC) [1, 6]. AHC (MIM 300220, http://www.ncbi.nlm.nih.gov/omim) is an inherited disorder of adrenal gland development, characterized by marked underdevelopment or absence of the permanent zone of the adrenal cortex and by structural disorganization of both the fetal cortex and the adult glands. The disorder, which is lethal if untreated, results in adrenal insufficiency early in infancy, with low serum concentrations of glucocorticoids, mineralcorticoids and androgens, and failure to respond to ACTH stimulation. Hypogonadotropic hypogonadism (HH) is commonly associated with the X-linked form of the disease and is generally diagnosed at the expected time of puberty. Clinical investigations of AHC patients provided conflicting results regarding hypothalamic or pituitary origin of the etiology of HH [7]. Expression studies have demonstrated that *DAX-1* messenger RNA (mRNA) is present in both the pituitary gland and in the hypothalamus (see below), consistent with a direct role for DAX-1 at both levels.

Many deletions and point mutations in the *DAX-1* gene have been described in AHC patients (reviewed in [8, 9]). Most point mutations (for updates, see the Human Gene Mutation Database, http://www.uwcm.ac.uk/uwcm/mg/hgmdO.html) are frameshift or nonsense mutations that introduce a premature stop codon and are distributed throughout the whole length of the transcript. All the mutations that cause single amino acid changes (9 missense and a single one-codon deletion) alter conserved amino acids in the ligand binding domain (LBD, see Fig. 1). Zhang et al. [8] have constructed a homology model of the DAX-1 LBD by homology to the three-dimensional crystal structure of the thyroid and retinoid X nuclear hormone receptors. The missence mutations and the codon deletion in DAX-1 all mapped to the predicted hydrophobic core of the LBD.

Clinical heterogeneity has been described among AHC patients, with some requiring hormonal replacement therapy within the first weeks after birth and others not being diagnosed until after 5 years of age. The variation in clinical presentation does not correlate with the extent of the molecular lesion and has even been described between members of the same AHC family. AHC clinical heterogeneity suggested that *DAX-1* mutations could be one of the causes of idiopathic HH or puberty delay. However, direct sequencing did not reveal any deleterious mutation in 106 patients [27].

Dax-1 functional absence in XY mice. A targeted disruption of *Dax-1* has recently been achieved by Yu et al. [10], using a conditional Cre-loxP approach. This strategy was needed because *Dax-1* was found to be essential for embryonic stem cell growth and differentiation. The disrupted allele lacks the second exon and mimics mutations found in AHC patients, Cre-mediated exision of the second exon also ablates the intronic splice acceptor site and the polyadenylation signal, resulting in low levels of an abnormal, unspliced transcript [10].

Unlike human AHC patients, Dax-1-deficient XY mice (AHC mice) are indistinguishable from their wild-type littermates until sexual maturation. Serum corticosterone levels in mutants are similarto those of wild-type animals. Histological inspection of the adrenals suggests that Dax-1 function is required for the initiation of fetal adrenal degeneration (which occurs after puberty), but is not necessary either for the formation of the definitive cortex or for steroidogenesis. Testosterone production during embryonic and early postnatal development is sufficient for the formation of male internal and external genitalia, for testicular descent (most AHC human patients show cryptochidism) and for the normal development of the testosterone-sensitive seminal vesicles. AHC mice are hypogonadal, but unlike AHC human patients their gonadotropin levels are normal, suggesting primary testicular failure. Testicular weights are approximately one half of those of wildtype animals. Dax-1 deficiency causes progressive degeneration of the testicular germinal epithelium and results in sterility.

DAX-1 functional absence in XX individuals. As expected in X-linked recessive disorders, heterozygous carriers of AHC mutations are generally normal, although skewed X inactivation might lead to the expression of a partial phenotype. Delayed puberty in female carriers has been described in one AHC family [28].

The occurrence of homozygous *DAX-1*-deficient human females is unlikely because males, who would be needed to transmit the non-functional allele, are sterile. An unusual AHC family, segregating a nonsense mutation in *DAX-1*, has been recently described. The unaffected maternal grandfather of two boys with a complete AHC syndrome was found to carry the mutation. In addition, the maternal aunt, who was affected by isolated HH, was found to be homozygous for the mutation with no mosaicism for the normal allele in three different tissues. Homozygosity was suggested to have been introduced by gene conversion early in embryogenesis [29].

Dax-1 functional absence in XX mice. The disruption of *Dax-1* in female mice does not appear to prevent sexual maturation, ovulation or fertility [10]. Histological analyses of ovaries from mature females show a normal complement of follicles at different stages of maturation a well as the presence of corpora lutea. Dax-1-deficient females mated with wild-tpye males produce normal litter sizes with equal transmission of the mutation to male and female offspring. The only abnormality found in mutant females is the presence of multiple oocytes in a subset of follicles. A single thecal layer is present and surrounds the proliferating granulosa cell layer and oocytes, implicating Dax-1 in follicular recruitment, granulosa cell proliferation or in the formation of structures that segregate different follicles.

Species differences in DAX-1 function

The comparison of natural human mutations and murine models reveals several apparent differences in the effect of DAX-1 dosage in humans and mice. A possible explanation of the dichotomy is that the murine models do not reflect accurately the physiological situation. The promoter sequences used for transgenic overexpression experiments could lack specific sequences required for precise timing of expression. In the targeted disruption experiments, the disrupted allele may encode a protein with residual biological activity. Alternatively, the observed species differences may reflect species-specific changes in the processes that control adrenal and gonadal development and function. Accordingly, the high degree of divergence between human and mouse DAX-1 proteins (see below) may suggest that they act in the context of a rapidly evolving system.

DAX-1 structure

The *DAX-1* gene encodes an unusual member of the nuclear hormone receptor (NR) superfamily. The protein is between 470 and 474 amino acids in length in the different mammalian species analysed to date [1, 11–14]. The DAX-1 protein can be divided into two parts with different structural and functional features (Fig. 1).

The carboxy-terminal half of the protein is similar to the ligand binding domain (LBD) of the NR superfamily. The highest homology is observed with SHP (an unusual member of the NR family that contains a canonical LBD but no DBD) [15], with the RXR and the orphan (see [1] for references) receptor subfamilies. The level of homology between DAX-1 and the other embers of the NR superfamily, although significant, does not indicate the nature of the hypothetical ligand. A remarkable feature of the human DAX-1 LBD is the presence of an unusually long insertion which is in the

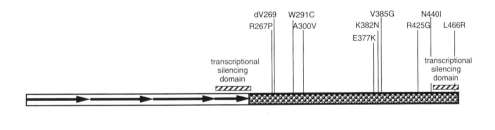

Figure 1. Schematic representation of the DAX-1 protein showing the putative DNA binding and ligand binding domains (DBD and LBD, respectively) and the transcriptional silencing domains. The AHC mutations that cause a single amino acid change are also shown.

proximity of the predicted ligand binding pocket. This region is poorly conserved between species and may imply the absence of a specific ligand for DAX-1 (Fig. 1). The most conserved portions of the LBD contain the transcriptional silencing domains (see below).

The DNA binding domain (DBD) of RN proteins is composed of two zinc fingers and is conserved amongst the different types of receptors. In DAX-1 the canonical DBD of NR proteins is missing and is substituted by a novel N-terminal repetitive structure, organized into four incomplete repetitions of an alanine- and glycinerich 65–67-amino acid motif. The repeats, showing 33–70% identity to each other, contain conserved cysteine residues. This N-terminal structure of DAX-1 shows no obvious similarity to previously reported protein sequences, and might define a novel DNA binding domain with specificity for DNA hairpin structures [16].

Lalli et al. [17] have compared the structure of the LBD of human and mouse DAX-1 to the three-dimensional structure of the LBD of apo-RXRα and holo-RARγ. They identified in DAX-1 the domains corresponding to α-helices 1–12, which represent the hallmark of the NR LBD. Their analysis also suggested that helix H1 of the DAX-1 LBD encompasses the region previously defined as the last and incomplete DBD repeat.

NRs are highly conserved between different species, with LBDs showing continuous similarity levels higher than 85%. Detailed comparison of human and mouse DAX-1 sequences [12] reveals subdomains where the predicted proteins are highly conserved (81–96% identity and 90–100% similarity) separated by amino acid stretches, accounting for approximately half of the protein, which are only poorly conserved (41–55% identity and 58–71% similarity). The protein-coding sequences of *DAX-1* and *Dax-1* are more similar at the DNA than at the protein level, suggesting rapid evolution. This pattern of evolution might be accounted for either by a process of random fixation of neutral point mutations in subregions lacking functional constraints or by positive selection for divergence. Positive selection is suggested by the very high nonsynonymous to synonymous substitution ($K_a/K_s > 0.8$) ratio in the central portion of the LBD. It is worth noting that *SRY* is also a rapidly evolving gene, whereas other genes involved in sex determination and gonadal differentiation such as *SOX9*, *WT1* and *SF-1* (see accompanying reviews) are highly conserved between mouse and humans. This type of rapid evolution could reflect competition between SRY and DAX-1 and be related to their positions at the top of a hierarchy.

DAX-1 pattern of expression

The phenotypic similarities that accompany disruption of the *Sf-1* gene (see accompanying review) and mutations in *DAX-1* suggest that these two genes act in the same developmental pathways, and *in situ* hybridization studies suggest a striking colocalization of the mRNAs for *Dax-1* and *Sf-1* within four developing organs: adrenals, gonads, hypothalamus and pituitary gland [12, 18, 19]. Together these results are consistent with DAX-1 and Sf-1 determining adrenal and gonadal development and modulating reproductive function at hypothalamic, pituitary and gonadal levels.

Dax-1 is expressed in the adrenals from the very earliest stages of their development and may be expressed within the cells making up the lineage that gives rise to adrenal cells prior to their separation from the genital ridge. The gene is then expressed at all stages of adrenal development including the adult organ. *Dax-1* expression is always restricted to the adrenal cortex [12, 19].

AHC human patients, but not *Dax-1* knockout mice, are also characterized by HH. *Dax-1* is expressed in the developing hypothalamus, and in the pituitary at low levels, which implies that the HH phenotype seen in AHC patients may be due directly to DAX-1 functional deficiency [12, 19]. By 11.5 dpc, *Dax-1* expression is detected in the same region of the developing brain that expresses *Sf-1*. By 14.5 dpc, expression has localized to the retrochiasmatic ventral diencephalon, which ultimately contributes to the hypothalamus, and by 18.5 dpc to the ventromedial hypothalamic nucleus, the site were both *Sf-1 and Dax-1* are expressed in the newborn and adult mouse brain [19]. The anterior pituitary gland contains five discrete cell types, of which only the gonadotropes express *Sf-1* and *Dax-1* [19].

The onset of *Dax-1* expression in both XX and XY genital ridges occurs at the same time as *Sry* in XY genital ridges. After 12.5 dpc, when *Sry* expression disappears, *Dax-1* expression in the testis decreases dramatically while it persists in the ovary. This supports the proposal that *Dax-1* interferes with testis development and that its expression must be repressed in testis.

Dax-1 positive cells show a distribution within the developing gonads that has not been observed previously. One interpretation of this pattern is that *Dax-1* marks a specific stage of relatively undifferentiated cells. These cells disappear rapidly in the testis, but remain longer in the ovary, especially in a region adjacent to the mesonephros. This distribution pattern could be used to argue that Sry initiates differentiation of the genital ridge into a testis which in turn represses *Dax-1* expression. The group of cells in the anterior region of the gonad expressing *Dax-1* in both sexes could correspond either to cells which have escaped the signal to differentiate or to a remnant of adrenal precursors [5, 12].

In the adult mouse, *Dax-1* is primarily expressed in steroidogenic tissues, including the adrenal cortex, testicular Leydig cells and ovarian thecal and granulosa cells [12, 19]. Expression in rat Sertoli cells was reported by Tamai et al. [20], who first suggested that Dax-1 may influence the development of spermatogenic cells in response to steroid and pituitary hormones. The expression is regulated during spermatogenesis and peaks during the androgen-sensitive phase of the spermatogenic cycle. Maximum levels are present between postnatal days 20 and 30 in the rat, during the first spermatogenic wave. Treatment of culture Sertoli cells with follicle stimulating hormone (FSH), an important regulator of spermatogenesis, results in a potent down-regulation of *Dax-1* expression [20]. These data fit with the recent finding that Dax-1-deficient male mice are sterile due to progressive epithelial dysgenesis with complete loss of germ cells by 14 weeks [10].

Regulation of the expression of *DAX-1*

Do SF-1 or WT1 regulate DAX-1?

The overlapping tissue distributions and known roles of SF-1 and DAX-1 raise the possibility that they interact either directly or indirectly. An obvious possibility is that the receptors act in a developmental cascade and one factor regulates the expression of the other gene. The 5′-flanking region of the human and mouse *DAX-1* genes contain a sequence that matches the consensus DNA binding motif for Sf-1 protein; this sequence binds SF-1 in gel mobility shift assays (EMSA) [19, 21]. However, 5′ deletion analyses were used to show that the putative SF-1-responsive element does not regulate expression of *Dax-1* either in mouse Y-1 adrenocortical cells (a cell line that does not express *Dax-1* endogenously) or in *Dax-1*-expressing MA-10 Leydig [19]. In addition, *Dax-1* expression persists in the gonads and the hypothalamus of a mouse with a disrupted *Sf-1* gene [19]. As suggested by Ikeda et al. [19], these findings imply that Sf-1 is not an obligatory positive regulator of *Dax-1*. However, significant impairment of *Dax-1* expression was more recently found in another mouse model with a disrupted *Sf-1* gene [30].

In a set of experiments, Yu et al. [22] discovered that the murine *Dax-1* promoter contains a duplicated binding site for Sf-1. These sites are not completely conserved in the human gene. The combined sites (-134 to -114 in the murine promoter) function a a composite element that is capable of interacting with several proteins in the nuclear extracts from αT3, a gonadotrope cell line that expresses both *Dax-1* and *Sf-1*, and Y-1, an adrenocortical cell line that expresses *Sf-1* but not *Dax-1*. In both cell lines, the formation of three major complexes was observed, and super-shift analyses identified two of these complexes as containing Sf-1 and COUP-TF. Both half sites were required for full Sf-1-mediated activation: disruption of these sites reduced by half the basal activity of the native *Dax-1* promoter in both αT3 and Y-1 cells. Disruption of the first site increased basal promoter activity in a placental choriocarcinoma cell line, JEG-3, that does not usually express *Dax-1* and *Sf-1*. These results suggest the existence of a repressor protein capable of interacting at this site. When linked to the minimal thymidine kinase promoter, each of the isolated Sf-1 sites was sufficient to mediate transcriptional regulation by Sf-1. Yu et al. [22] proposed that *Dax-1* expression is stimulated by Sf-1, and that Sf-1 and COUP-TF provide antagonistic pathways that converge upon a common regulatory site. Activation of *Dax-1* expression by WT1 has recently been described. Footprinting analysis, transient transfections, promoter mutagenesis and mobility shift assays all suggested that WT1 regulates *Dax-1* via GC-rich binding sites in the *Dax-1* promoter [31].

Does DAX-1 repress its own expression?

The observation that Dax-1 can bind to hairpin DNA structures (see below) prompted a search for such structures in gene promoters likely to be under Dax-1 regulation. Zazopoulos et al. [16] detected two possible hairpins (H1 and H2) in the promoter of the murine *Dax-1* gene and showed that H1 formation is favoured over H2. In transient transfection experiments, in COS-1 cells, using a reporter gene carrying the luciferase gene under the control of *Dax-1* promoter sequences, Dax-1 protein moderately represses the basal activity of both wild-type and H1-deleted promoters, whereas it drastically blocks Sf-1-mediated activation of the wild-type but not H1-delated *Dax-1* promoter. It was suggested that the proximity of Sf-1 and Dax-1 binding sites in the *Dax-1* promoters may allow an allosteric inhibition of Sf-1 binding (see below) [16].

The promoter defined by transgenic experiments

An 11-kb genomic fragment, derived from the region immediately upstream of the *Dax-1* start of transcription, is capable of targeting expression of a *β*-galactosidase reporter gene in the developing gonads in a pattern identical to the endogenous *Dax-1* gene [5]. In the transgenic animals, *β*-galactosidase activity was first observed at 10.5 dpc within the genital ridge, and reached its highest level by 11.5 dpc in both male and female embryos. This activity persists throughout ovary formation, but in the male it is rapidly downregulated around 12.5 dpc, although it is maintained in the region between the gonad and mesonephros in a manner identical to the endogenous gene. In the adult animals, reporter gene expression also followed that of the endogenous gene in gonadal sites. Interestingly, *β*-galactosidase activity was absent from other sites of endogenous *Dax-1* expression, such as the developing adrenal, hypothalamus and pituitary, suggesting that additional upstream sequences are necessary to obtain the complete pattern of expression of this gene [5].

DAX-1 acts as a potent transcriptional repressor

Several groups have reported that DAX-1 acts, *in vitro* and *in vivo*, as a potent transcriptional repressor. DAX-1 inhibits SF-1-mediated trans-activation of target genes and antagonizes SF-1/WT1 synergy – probably through a direct interaction with SF-1. Dax-1 also represses *StAR* (steroid-ogenic acute regulatory protein) gene expression.

DNA-binding activity of DAX-1

Zazopoulos et al. [16] have found that DAX-1 efficiently recognizes DNA hairpin structures *in vitro*. Binding is equally efficient to stems composed of 10–24 nucleotides, but is less efficient with shorter stems. The sequence of the loop influences DAX-1 binding: an adenine-rich sequence induces reduced binding compared with loops rich in cytosine or thymine. Methyla-tion interference and distamycin-binding experiments indicate that DAX-1 predominantly interacts with the minor groove of the DNA helix. Za-zopoulos et al. [16] have noted that these DNA-binding properties are reminiscent of high mobility group (HMG)-box proteins, such as SRY. However, whereas HMG proteins can bind to four-way DNA junctions, DAX-1 is unable to do so if the loop is missing. It is predicted that this new DNA-binding activity is a feature of the DAX-1 putative DBD. This domain has also been involved in protein-protein interaction (see below).

Does DAX-1 regulate expression of the StAR gene?

The spatial and temporal pattern of *DAX-1* expression in the adrenal cortex and Leydig cells suggested that it could be involved in the regulation of steroidogenesis. The steroidogenic acute regulatory protein (StAR) has a central role in steroidogenesis. Zazopoulos et al. [16] identified a DNA hairpin structure in the promoter of the *StAR* gene and demonstrated that it binds DAX-1. Transcription of the human *StAR* gene is induced by cyclic AMP (cAMP). DAX-1 represses both the basal expression and cAMP-induced activity of a *StAR* promoter reporter construct in the Y-1 cell line. A *StAR* promoter mutated in the hairpin structure is still cAMP-inducible, but is not repressed by DAX-1. In addition, *StAR* transcripts and protein are undetectable even after forskolin stimulation in Y-1 clones expressing DAX-1 after transfection [16]. Experiments using a hydroxylated chol-esterol derivative show that biochemical steps in steroidogenesis sub-sequent to cholesterol delivery to mitochondria are also impaired in Y-1 cells expressing DAX-1. This is explained by the repression of P450scc and 3β-HSD expression, in addition to StAR. DAX-1 expression in Y-1 cells results in the inhibition of the activity of the StAR, P450scc and 3β-HSD promoters [23]. This suggests a prominent role for DAX-1 in the control of steroidogenesis. However, examination of a variety of adernocortical tumours failed to detect the expected negative correlation between *DAX-1* and *StAR* expression [24].

Interaction between DAX-1 and SF-1 at the protein level

Interaction between SF-1 and DAX-1 proteins was demonstrated by Ito et al. [25] in a transient transfection model using the JEG-3 cell line. Transfection of an *SF-1* expression construct activates a reporter gene containing one or two copies of the SF-1 binding site. Cotransfection with a *DAX-1* expression construct reduces expression of the reporter gene. However, DAX-1 protein does not bind directly to the SF-1 site in gelshift assays, nor does it alter the binding of SF-1 to its response element. In addition, cotransfection of *DAX-1* and *SF-1* did not apparently result in heterodimer formation; however, protein-protein interaction was detected in *in vitro* protein binding assays. DAX-1 C-terminal deletions (between amino acids 470 to 488) or the two naturally occurring amino acid substitutions R267P and DV269 failed to reduce DAX-1 binding to SF-1, whereas removal of the N-terminal region (amino acids 1 to 226) decreased binding.

A direct *in vitro* interaction between DAX-1 and SF-1 in the absence of DNA was confirmed (but not published) by Zazopoulos et al. [16], who were unable to demonstrate interaction *in vivo*. They proposed that repression by DAX-1 in an natural promoter context might require DNA binding.

More recently, Nachtigal et al. [26] showed that SF-1 acts synergistically with WT1 to promote anti-Müllerian hormone (*MIS*) expression (see Parker et al., this issue). This activation could be blocked by DAX-1. Removing both DAX-1 silencing domains (see below) was required to relieve the repression. Cotransfection of *DAX-1* and *SF-1* in the absence of WT1 did not cause a similar inhibitory effect. In a yeast two-hybrid assay, a prominent association between SF-1 and DAX-1, but not WT1 and DAX-1, was detected. Furthermore, binding between DAX-1 and SF-1 was also detected in a GST pulldown experiment.

Removal of the SF-1 LBD abrogated its interaction with GST-DAX-1, but not with GST-WT1, suggesting that the major sites conferring interaction with DAX-1 and WT1 are mediated by different regions of SF-1. Nachtigal et al. [26] suggested that failure by other authors to detect a direct interaction *in vivo* might be due to the use of human DAX-1 vs. mouse Sf-1, possible due to the high degree of protein divergence between human and mouse DAX-1. Nachtigal et al. [26] also confirmed that DAX-1 is unable to bind or form a visible protein-protein complex on a SF-1 responsive element (MIS-RE-1).

The transcriptional silencing domains of DAX-1

The domains essential for transcriptional silencing were identified by the creation of a series of deletions and mutations within the C-terminus of DAX-1. These were fused in frame to the yeast GAL4 DBD [17, 25]. The effect of these fusion proteins on transcription driven by different promoters led to the identification of two silencing domains: removal of the most C-terminal 19 [16, 17] or 28 [25] amino acids results in almost complete abrogation of silencing. Lalli et al. [17] also examined the N-terminal portion of DAX-1 and demonstrated that deletions of amino acids 207–244 significantly reduces silencing in one of the systems used (Fig. 1).

References

1 Zanaria E, Muscatelli F, Bardoni B, Strom TM, Guioli S, Guo W et al (1994) An unusual member of the nuclear hormone receptor superfamily responsible for X-linked adrenal hypoplasia congenita. *Nature* 372: 635–641

2 Bardoni B, Zanaria E, Guioli S, Floridia G, Worley KC, Tonini G et al (1994) A dosage sensitive locus at chromosome Xp21 is involved in male to female sex reversal. *Nature Genet* 7: 497–501

3 Muscatelli F, Walker AP, De Plaen E, Stafford AN and Monaco AP (1995) Isolation and characterization of a MAGE gene family in the Xp21.3 region. *Proc Natl Acad Sci USA* 92: 4987–4991

4 Dabovic B, Zanaria E, Bardoni B, Lisa A, Bordignon C, Russo V et al (1995) A family of rapidly evolving genes from the sex reversal critical regionin Xp21. *Mamm Genome* 6: 571–580

5 Swain A, Narvaez V, Burgoyne P, Camerino G and Lovell-Badge R (1998) *Dax1* antagonizes *Sry* action in mammalian sex determination. *Nature* 391: 761–767

6 Muscatelli F, Strom TM, Walker AP, Zanaria E, Recan D, Meindl A et al (1994) Mutations in the *DAX-1* gene give rise to both X-linked adrenal hypoplasia congenita and hypogonadotropic hypogonadism. *Nature* 372: 672–676

7 Kletter GB, Gorski JL and Kelch RP (1991) Congenital adrenal hypoplasia and isolated gonadotropin deficiency. *Trends Endocrinol Metab* 2: 123–128

8 Zhang YH, Guo W, Wagner RL, Huang BL, McCabe L, Vilain E et al (1998) *DAX1* mutations map to putative structural domains in a deduced three-dimensional model. *Am J Hum Genet* 62: 855–864

9 Peter M, Viemann M, Partsch CJ and Sippell WG (1998) Congenital adrenal hypoplasia: clinical spectrum, experience with hormonal diagnosis and report on new point mutations of the *DAX-1* gene. *J Clin Endocrinol Metab* 83: 2666–2674

10 Yu RN, Ito M, Saunders TL, Camper SA and Jameson JL (1998) Role of *Ahch* in gonadal development and gametogenesis. *Nature Genet* 20: 353–357

11 Guo W, Burris TP, Zhang YH, Huang BL, Mason J, Copeland KC et al (1996) Genomic sequence of the *DAX1* gene: an orphan nuclear receptor responsible for X-linked adrenal hypoplasia congenita and hypogonadotropic hypogonadism. *J Clin Endocrinol Metab* 81: 2481–2486

12 Swain A, Zanaria E, Hacker A, Lovell-Badge R and Camerino G (1996) Mouse *Dax1* expression is consistent with a role in sex determination as well as in adrenal and hypothalamus function. *Nature Genet* 12: 404–409

13 Guo W, Lovell RS, Zhang YH, Huang BL, Burris TP, Craigen WJ et al (1996) *Ahch*, the mouse homologue of *DAX1*: cloning, characterization and synteny with *GyK*, the glycerol kinase locus. *Gene* 178: 31–34

14 Parma P, Pailhoux E, Puissant C and Cotinot C (1997) Porcine *Dax-1* gene: isolation and expression during gonadal development. *Mol Cell Endocrinol* 135: 49–58

15 Seol W, Choi HS and Moore DD (1996) An orphan nuclear hormone receptor that lacks a DNA binding domain and heterodimerizes with other receptors. *Science* 272: 1336–1339

16 Zazopoulos E, Lalli E, Stocco DM and Sassone-Corsi P (1997) DNA binding and transcriptional repression by DAX-1 blocks steroidogenesis. *Nature* 390: 311–315

17 Lalli E, Bardoni B, Zazopoulos E, Wurtz JM, Strom TM, Moras D et al (1997) A transcriptional silencing domain in DAX-1 whose mutation causes adrenal hypoplasia congenita. *Mol Endocrinol* 11: 1950–1960

18 Gua W, Burris TP and McCabe ER (1995) Expression of *DAX-1*, the gene responsible for X-linked adrenal hypoplasia congenita and hypogonadotropic hypogonadism, in the hypothalamic-pituitary-adrenal/gonadal axis. *Biochem Mol Med* 56: 8–13

19 Ikeda Y, Swain A, Weber TJ, Hentges KE, Zanaria E, Lalli E et al (1996) Steroidogenic factor 1 and Dax-1 colocalize in multiple cell lineages: potential links in endocrine development. *Mol Endocrinol* 10: 1261–1272

20 Tamai KT, Monaco L, Alastalo TP, Lalli E, Parvinen M and Sassone-Corsi P (1996) Hormonal and developmental regulation of *DAX-1* expression in Sertoli cells. *Mol Endocrinol* 10: 1561–1569

21 Burris TP, Guo W and McCabe ER (1995) Identification of a putative steroidogenic factor-1 response element in the *DAX-1* promoter. *Biochem Biophys Res Commun* 214: 576–581

22 Yu RN, Ito M and Jameson JL (1998) The murine *Dax-1* promoter is stimulated by SF-1 (steroidogenic factor-1) and inhibited by COUP-TF (chicken ovalbumin upstream promoter-transcription factor) via a composite nuclear receptor-regulatory element. *Mol Endocrinol* 12: 1010–1022

23 Lalli E, Melner MH, Stocco DM and Sassone-Corsi P (1998) DAX-1 blocks steroid production at multiple levels. *Endocrinology* 139: 4237–4243

24 Reincke M, Beuschlein F, Lalli E, Arlt W, Vay S, Sassone-Corsi P et al (1998) DAX-1 expression in human adrenocortical neoplasms: implications for steroidogenesis. *J Clin Endocrinol Metab* 83: 2597–2600

25 Ito M, Yu R and Jameson JL (1997) DAX-1 inhibits SF-1-mediated transactivation via a carboxy-terminal domain that is deleted in adrenal hypoplasia. *Mol Cell Biol* 17: 1476–1483

26 Nachtigal MW, Hirokawa Y, Enyeart-VanHouten DL, Flanagan JN, Hammer GD and Ingraham HA (1998) Wilms' tumor 1 and Dax-1 modulate the orphan nuclear receptor SF-1 in sex-specific gene expression. *Cell* 93: 445–454

27 Achermann JC, Gu WX, Kotlar TJ, Meeks JJ, Sabacan LP, Seminara SB et al (1999) Mutational analysis of DAX1 in patients with hypogonadotropic hypogonadism or pubertal delay. *J Clin Endocrinol Metab* 84: 4497–4500

28 Seminara SB, Achermann JC, Genel M, Jameson JL and Crowley WF Jr (1999) X-linked adrenal hypoplasia congenita: a mutation in DAX1 expands the phenotypic spectrum in males and females. *J Clin Endocrinol Metab* 84: 4501–4509

29 Merke DP, Tajima T, Baron J and Cutler GB Jr (1999) Hypogonadotropic hypogonadism in a female caused by an X-linked recessive mutation in the DAX1 gene. *N Engl J Med* 340: 1248–1252

30 Kawabe K, Shikayama T, Tsuboi H, Oka S, Oba K, Yanase T et al (1999) Dax-1 as one of the target genes of Ad4BP/SF-1. *Mol Endocrinol* 13: 1267–1284

31 Kim J, Prawitt D, Bardeesy N, Torban E, Vicaner C, Goodyer P et al (1999) The Wilms' tumor suppressor gene (wt1) product regulates Dax-1 gene expression during gonadal differentiation. *Mol Cell Biol* 19: 2289–2299

Genes and Mechanisms in Vertebrate Sex Determination
ed. by G. Scherer and M. Schmid
© 2001 Birkhäuser Verlag Basel/Switzerland

Sex chromosomes and sex-determining genes: insights from marsupials and monotremes

Andrew Pask[1] and Jennifer A. Marshall Graves[2]

[1] Department of Zoology, The University of Melbourne, Parkville, Vic 3052, Australia
[2] School of Genetics and Evolution, La Trobe University, Melbourne, Vic 3083, Australia

Summary. Comparative studies of the genes involved in sex determination in the three extant classes of mammals, and other vertebrates, has allowed us to identify genes that are highly conserved in vertebrate sex determination and those that have recently evolved roles in one lineage. Analysis of the conservation and function of candidate sex determining genes in marsupials and monotremes has been crucial to our understanding of their function and positioning in a conserved mammalian sex-determining pathway, as well as their evolution. Here we review comparisons between genes in the sex-determining pathway in different vertebrates, and ask how these comparisons affect our views on the role of each gene in vertebrate sex determination.

Introduction

Sex determination is critical for reproduction. One would think, therefore, that the mechanisms determining sex and sexual dimorphisms would be extremely conserved in evolution. Certainly the process of gonadogenesis appears to be similar, at least at the histological level, but, surprisingly, the control of this critical phenotype seems to be subject to great variation within mammals, and between mammals and other vertebrates. We can use this variation to ask how the sex determining pathway evolved, and how it functions.

Particularly enlightening have been comparisons between the three major extant groups of mammals. Marsupials and monotremes represent the mammals most distantly related to humans and mice, having diverged from eutherians about 130 and 170 million years before present (MYrBP) respectively [1], early in the 200 million year history of Class Mammalia. In turn mammals diverged from reptiles and birds about 350 MYrBP (Fig. 1).

Sex determination in mammals is accomplished by a chromosomal mechanism. Females (the homogametic sex) have two X chromosomes, and males (the heterogametic sex) a single X and a Y. In eutherian ("placental") mammals, observations of the phenotypes of XO females and XXY males shows that the presence of a Y chromosome determines a male phenotype, no matter how many copies of the X are present. This was ascribed to the presence on the Y of a dominant "testis determining factor" (TDF), which activates a testis determining pathway. Once a testis is differentiated in the embryo, the hormones it produces control all other aspects of male phenotype.

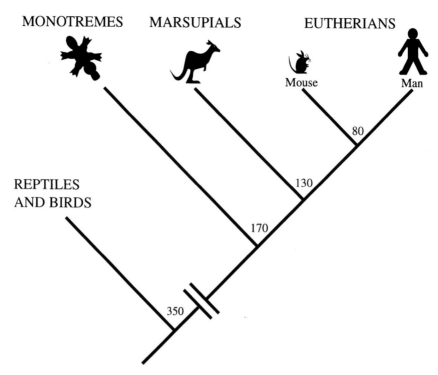

Figure 1. Phylogeny of class Mammalia, indicating the divergence of marsupials, monotremes, reptiles and birds. Numbers indicate estimated times of divergence in millions of years before present.

In marsupials, the control of testis determination is also vested in the Y, for XXY animals have testes and XO do not. However, not all aspects of male phenotype are fixed by testicular androgens [2]. XXY animals have no scrotum, but possess a pouch with mammary glands. XO, though more variable, are the reverse, lacking pouch and mammary glands, but having a scrotum. This suggests that there is a gene on the marsupial X which controls a scrotum/mammary switch. This gene must be either dosage sensitive such that one copy determines scrotum development and two mammary development, or imprinted such that gene(s) on the paternal X switches the potentiality from scrotum to mammary gland development [3].

There are several different modes of sex determination in other vertebrates. Birds, as well as snakes subscribe to a ZZ male : ZW female sex chromosome system in which the female, not the male, is the heterogametic sex, and the heterochromatic W chromosome is weakly female-determining. Reptiles have a great variety of chromosomal and genetic sex determination mechanisms, as well as environmental sex determination, in which a factor such as temperature of incubation determines the sex of the eggs. Do all these control mechanisms operate through the same genetic pathway? Are the same sex determining genes active in testis determination

– and ovary determination – in all vertebrates? Here we will review the evidence for the identity of genes in the sex determining pathway in eutherian mammals, and ask whether the same genes operate in the same pathway in other mammal groups and other vertebrates. Comparisons of homologous genes between closely and distantly related species have provided some surprising information on the complexity and variation of the controls on sex determination, and have demonstrated that a knowledge of evolution can sometimes be required for an understanding of function.

Sex chromosome organization, function and evolution

In mammals, the X and Y chromosomes are strikingly non-homologous, having different sizes and quite different gene contents. The human X chromosome comprises about 5% of the haploid genome, and contains 3000 or 4000 genes. These code for a mixture of classic housekeeping enzymes and products with specialised functions, although it has been argued that there are a disproportionate number of X-linked conditions with an effect on gonads or reproduction. The X is highly conserved between different eutherian species, perhaps because of a chromosome-wide inactivation mechanism which ensures dosage compensation between males and females.

The human Y chromosome is much smaller and largely heterochromatic. It recombines with the X only over a tiny "pseudoautosomal region" (PAR) at the tips of the short arms, and a second smaller homologous region at the tips of the long arms. Only three phenotypes were initially ascribed to the Y: the testis determining factor, TDF; a minor male-specific antigen HYA; and a region AZF whose deletion confers sterility in azoospermic men. However, several genes and pseudogenes have been isolated from cloned regions of the Y, or by homologous cloning using probes which map elsewhere on the genome [4]. Many sequences on the differential part of the Y detect a homologue on the X chromosome. Unlike the X, the Y is poorly conserved between species, and there are several genes which are active in one species and inactive in another. The gene content of the PAR is not conserved, but represents different subsets of markers present on human Xp [5]. The content and activity of genes on the Y chromosome has therefore changed rapidly during recent eutherian evolution.

A useful way to unravel ancient events in the evolution of the mammalian sex chromosomes has been to look for variation of sex chromosomes among the three major extant groups of mammals. The basic marsupial X is smaller than the eutherian X, and the basic Y is tiny (estimated at 12 Mb). The marsupial X and Y seem not to undergo homologous pairing at meiosis [6], and presumably have no pseudoautosomal region. In contrast, the monotreme sex chromosomes are large, and the X and Y pair at meiosis over the entire short arm of the X and long arm of the Y [7].

In order to detect genetic homology between the sex chromosomes of these three groups, comparative mapping has been undertaken (reviewed [8]). Somatic cell genetics and in situ hybridization have shown that the X chromosome of marsupials includes all the genes on the long arm and the pericentric region of the human X. The same suite of genes lies on the monotreme X, so must represent a highly conserved original mammalian X. However, the marsupial and monotreme X lack the genes on the rest of the short arm of the human X. Since marsupials and monotremes diverged independently from eutherians, the most parsimonious explanation is that this region was recently added to the X in the eutherian lineage. The observation that most genes within this region map to two similar clusters in marsupials and monotremes suggests that there have been at least two additions to the X (Fig. 2).

The marsupial Y chromosome shares at least four genes with the Y of humans and/or mice. *SRY*, *RBM1*, *SMCY* and *UBE1Y* have been cloned and mapped to the marsupial Y [9–12; Duffy, unpublished observations]. However, a number of genes within the recently added region which have homologues on the human X and Y (including genes on the human and mouse PAR) detect only autosomal sequences in marsupials [13]. This implies that the region was recently added, not only to the eutherian X, but also to the eutherian Y.

The observations that most or all of mammalian Y-borne genes have X-linked homologues, and that the eutherian X and Y share a homologous PAR, confirm that the X and Y evolved from a pair of autosomes in an ancestral mammal. This hypothesis was proposed long ago by analogy to snake sex chromosomes, which are thought to represent a series of states intermediate between undifferentiated and strongly differentiated Z and W chromosomes [14]. It receives support from comparative studies of mammal and bird sex chromosomes. Limited mapping of genes homologous between birds and mammals reveals none shared between the conserved mammal X and the bird Z. Mammalian X linked genes are scattered among bird autosomes, and most bird Z-linked genes map to human chromosome 9 [15, 74]. This implies that the two sex chromosome systems evolved independently from different autosomal pairs. It is not possible to deduce the sex determining system of the common reptilian ancestor, because other vertebrate classes show a wide variety of genetic and environmental sex determination.

It is thought that the original Y was first differentiated when an allele took on a dominant male-determining function. Other genes with a male-specific function, or conferring a male advantage, accumulated near it in a region of low recombination. Once this region became genetically isolated from its homologue, mutations, insertions and deletions accumulated [16]. However, this rapid degeneration of the Y was offset by cycles of addition to the eutherian X and Y. An "addition-attrition" hypothesis has been presented in which autosomal regions have been progressively added to the X

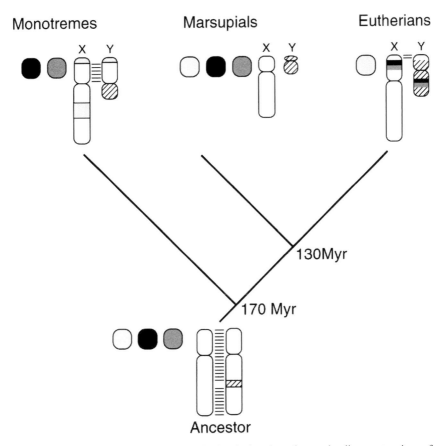

Figure 2. Addition/Attrition hypothesis. The hatched regions denote the divergent regions of the Y chromosome that began with the acquisition of the initial testis-determining gene on the proto Y chromosome in an ancestral mammal. Lines between the X and Y chromosomes denote regions of shared homology. The three shaded blocks in the ancestral mammal represent three different autosomal regions of the genome. Two of these autosomal blocks of genes have been added to the eutherian X and Y chromosomes, and one has been added to the monotreme X and Y chromosomes. Marsupials have not acquired any additional autosomal material to their X chromosome, and the X and Y are fully differentiated.

above the PAR, recombined onto the Y by exchange within the PAR, extending the PAR in stages [8] (Fig. 2). The initial addition could equally well have been to the Y, with recombination onto the X. One way or the other, the X and the Y have grown incrementally during eutherian, but not marsupial, evolution. Independent additions have been made to the X and Y in monotremes, and in one group of marsupials [5]. The autosomal regions added to the mammalian Y seem to have suffered a fate similar to that of the original Y, though more recently, so that proportionately more genes in the recently added regions of the X still have homologues on the Y.

The continued inactivation and loss of alleles from the Y chromosome would impose gene dosage differences between males and females were it not for the spread of inactivation along the X which more-or-less kept pace with Y attrition [17]. Exceptions are the several human X-Y shared genes, which escape from inactivation, and several genes whose activity states show that Y attrition and X inactivation is sometimes out of step.

Of the thousands of genes originally present on the proto-Y, only a few remain, and many of these are inactive or partially active copies of genes on the X. What selective forces operated to retain active genes on the Y, against this inexorable degradation? Conceivably, a gene might be so dosage sensitive that haploin sufficiency is deleterious or lethal. However, almost no X-Y shared genes outside the PAR share identical functions and activity. It seems more likely, then, that a Y-specific gene is retained for long periods because it acquires a unique function, presumably a male-specific one that is critical for male determination or differentiation. It seems certain, then, that genes originally with a general function in both sexes progressively took on male-specific functions in sex determination and male differentiation, while their X-linked partners retained the original ubiquitous function in both sexes.

Thus genes on the Y chromosome in different mammals seem to represent small, non-identical but overlapping subsets of genes on the X. Some, evidently expendable, genes seem to be dead or dying, while others appear to serve a male-specific function which ensures their survival over long periods of evolutionary time. The cloning and characterization of the *SRY* gene, and the investigation of related genes suggests that the *SRY* gene itself acquired its testis-determining function in this haphazard manner.

Comparative mapping in the hunt for the testis determining factor

The formation of a testis in the mammalian embryo is the primary sex-determining event, leading to the expression of male hormones, which in turn induce embryo masculinisation. In the absence of the Y chromosome, the embryo follows the female development pathway. The Y must therefore contain a testis determining factor (TDF for testis determining factor in humans and Tdy for testis determining gene on the Y in mouse). A positional cloning approach, relying on deletion analysis in human and mouse, was taken to pinpoint and clone the TDF gene. Comparative mapping and analysis of candidate genes in other species played a critical part in the correct identification of the mammalian TDF.

The first gene which was identified within a critical region was *ZFY* (*z*inc *f*inger protein on *Y* chromosome) [18], which encodes a putative DNA binding protein which could function as a transcriptional regulator and trigger a pathway for testis differentiation. *ZFY* mapped to the Y chromosome in all eutherian mammals tested, and detected a homologous sequence on the short arm of the X.

The first indication that *ZFY* was not the sex determining gene came with the finding that *ZFY* homologues mapped to autosomes in several marsupial species [19], colocalizing with other genes within the recently added region. As marsupial mammals have an XY sex determining mechanism and share at least part of the Y with eutherians, and the marsupial Y is testis determining (if not entirely responsible for the male phenotype), we would assume that the same TDF gene would be on the Y in marsupials as well as eutherian mammals. The autosomal location of *ZFY* in marsupials means that it could not act as a universal male determining switch gene, contradicting the hypothesis that *ZFY* could be a conserved TDF in therian (eutherian and marsupial) mammals. This conclusion was confirmed by expression studies [20] and by the observation of XX males that lack the *ZFY* gene [21]. Another gene or genes on the Y chromosome must therefore be responsible for sex determination.

The *SRY* gene (*sex determining region on the Y chromosome*) was isolated from the newly defined critical sex determining region on the human Y and detected male specific homologues in all eutherian mammals tested [22]. The human *SRY* gene encodes a protein that contains a 79 amino acid conserved high mobility group (HMG) box which can bind to and bend DNA. Mutation analysis showed that XY females have mutations within *SRY* [23]. *SRY* is a member of a large gene family of HMG box containing (*SOX*) genes.

Critical to acceptance of *SRY* as the mammalian testis-determining gene was the demonstration that the mouse homologue (*Sry*), mapped within the critical sex-determining region of the mouse Y chromosome [24]. Furthermore, *Sry* had an appropriate expression profile in the developing testis, and promoted male development of XX transgenic mice [25]. Finally, to everyone's relief, a marsupial *SRY* homologue was cloned and mapped to the Y chromosome [26]. In the absence of mutation analysis or the prospect of transgenesis, it has not been possible to confirm that *SRY* is testis determining in marsupials, but its long association with the mammalian testis determining chromosome has been confirmed.

Comparative analysis of SRY in other mammals

The testis-determining gene, thought to code for a transcription factor critical for reproduction, was expected to be highly conserved between species. It therefore came as a surprise to find that *SRY* sequences of human, mouse and marsupial species are poorly conserved even within the HMG box [27]. *SRY* is even interrupted by a *de novo* intron in one marsupial family [28], and has been amplified in several species of old-world mice.

Outside the HMG box, sequences cannot even be aligned. A potential transactivating domain in the 3′ region of the mouse SRY protein [29] is not conserved in human or marsupial SRY, although it has been recently shown to be essential for activity in mouse [30]. Similarly, the unique C-terminal domain of human SRY, that can bind nuclear proteins containing two PDZ domains, such as *SIP*1 [31] is absent from mouse or marsupial SRY. This suggests that all the conserved sex determining activity of SRY is in the HMG box, a conclusion reinforced by the finding that almost all of the known amino acid substitutions found in mutant SRY proteins from XY females are within this region [23].

Equally as puzzling are the inconsistent expression patterns of *SRY* in different species. In mouse, *Sry* is expressed appropriately in the developing mouse gonadal ridge in males at day 11.5 p.c. (post coitum), the time at which the first histological signs of testis differentiation are noted [32]. *Sry* is also expressed in the adult testis as a unique non-translated circular transcript [33]. Human *SRY* is transcribed at a low level in many embryonic tissues, but is limited to the testis in adults. The finding of XY females with a mutations 5′ to a normal *SRY* gene [34] suggests that 5′ regulatory elements control the time and location of *SRY* expression. In contrast to the restricted expression patterns in mouse and human, *SRY* is expressed virtually ubiquitously in marsupials. In the tammar wallaby, *SRY* is transcribed in the embryo at every stage sampled, as well as in a wide range of adult tissues [35]. The significance of *SRY* transcripts in developing tissues other than testis is unclear. Does *SRY* have a function other than testis determination in marsupials?

The action of SRY is thought to depend on the properties of the HMG box, which binds to a 6-bp target sequence, and bends DNA through a specific angle [36, 37], which may promote association of distant regulatory elements into a complex that can control the activity of other genes. The angle of bending is very specific, and recombinant products of some mutant human *SRY* genes are deficient in bending. It is puzzling, therefore, that the HMG box bends DNA through quite different angles in different species.

The mechanism by which *SRY* acts to initiate testis determination is elusive. Because of the positive action of the testis-determining factor in promoting male development, it has always been expected that the testis-determining gene *SRY* would prove to function by transcriptional activation of testis differentiating genes. Indeed, the products of other HMG-box-containing genes are transcriptional activators, and SRY does show transcriptional activity *in vitro* [38].

However, the testis determining factor could operate equally well by repressing a gene which overrides *SRY* [39]. An inhibitory role for *SRY* was previously suggested to account for the puzzling cases of XX males who lack *SRY* [40]. In fact, there are entire species of rodents that lack *SRY*. The mole voles *Ellobius lutescens* and *E. tancrei*, undergo apparently normal sex determination, although they lack a Y chromosome, and have no *SRY* gene homologue [41]. Evidently some other gene has taken over the primary sex determining function in triggering the male developmental pathway. These observations suggest that *SRY* acts only as a switch to initiate the male development pathway and contributes little to male differentiation itself.

Other genes must therefore act in a sex determining pathway. Comparative analysis has been helpful in assessing the credentials of candidate genes.

Comparative analysis of genes in the sex-determining pathway

The next step in understanding how the *SRY* gene functions in male determination was therefore to look for genes which lie up or downstream of *SRY* in the testis determining pathway. Three approaches have been taken; to search for proteins that interact with *SRY*, to examine genes which code for testis-specific products such as AMH and RBM, and genes which are involved in sex reversal syndromes such as *SOX*9, *DAX*1 and ATRX.

The search for sequences to which *SRY* directly binds, or proteins that interact directly with *SRY*, has been frustrating. The 6-bp target site for binding of the HMG box of *SRY* is present in many genes, and is shared with many other HMG box-containing proteins. The HMG box also contains a calmodulin binding domain [42], also shared with other HMG box-containing proteins. Calmodulin binding may facilitate a conformational change in SRY protein that affects its ability to bind to its target sequence and thereby regulates its activity. In addition *SIP*1 (*SRY interacting protein* 1) has been identified by its ability to bind and interact with the C-terminus of the human SRY protein [31], and may facilitate *SRY* binding or bending its target site.

Genes which code for testis-specific products are good candidates for target genes. AMH (anti-Müllerian hormone), the first substance secreted from the Sertoli cells in the developing testis, induces regression of the female Müllerian tubules in the male embryo [reviewed 43]. *AMH* is an obvious potential target for initiation by *SRY*. Indeed, the SRY protein has been shown to bind to the *AMH* promoter, suggesting that *SRY* acts as a transcriptional regulator of *AMH* [44]. However, the possession of the target site may be fortuitous, and *in vitro* transcription assays misleading. A direct interaction between *SRY* and *AMH* is unlikely in mouse, where *AMH* is expressed long after *SRY* has ceased transcription in the developing testis.

There is some evidence that *AMH* expression may actually be an early event, at least in other vertebrates, since transcription coincides with the earliest signs of histological differentiation in chicken [57]. It may be in a parallel pathway, or even have some upstream effect on testis determination, as suggested by freemartin cattle, where sex reversal in female twins is induced by *AMH* in shared circulation with a male twin embryo [45].

Study of sex reversal in human patients and mouse mutants has yielded genes that could be part of the mammalian sex-determining pathway. For instance, mutations in the *SF*1 (*steroidogenic factor* 1) gene cause complete gonadal dysgenesis in both sexes [46], placing it upstream of *SRY* in the sex determining pathway, in the formation of the indifferent gonad. *SF*1 encodes an orphan nuclear receptor which regulates steroid hydroxylases. In humans, it is expressed throughout the reproductive system, hypothalamus, pituitary, adrenal cortex and gonads. *SF*1 has been shown to regulate AMH directly [47], but seems to have no direct interaction with *SRY*.

Sex-reversal syndromes also provide candidate genes for a role in sex determination. *SOX*9 on human chromosome 17q was shown to be responsible for a bone and cartilage syndrome, campomelic dysplasia, associated with XY female sex reversal [49, 50, 51]. *SOX*9 appears to be a transcription factor, containing an HMG box, which can bind to the same consensus sequence as *SRY*, and a C-terminal transactivation domain [54]. *SOX*9 is highly conserved between human, mouse, and even chicken, alligator and fish homologues [52, 55–57]. Human *SOX*9 is highly expressed in the testis [49], and mouse *Sox*9 showed specific expression in the developing male, but not the female, gonadal ridge, being upregulated in the developing sex cords and the Sertoli cells at the time of gonadal differentiation [55, 58]. Its induction coincides with *Sry* expression, but *Sox*9 remains active in embryonic testis long after *Sry* expression has ceased. Thus *SRY* may act by turning on *SOX*9, or may act with *SOX*9 in the developing mammalian testis to bring about Sertoli cell differentiation, but is not required for *SOX*9 maintenance.

Despite fundamental differences in chromosomal sex determination in birds, in which females are the heterogametic sex, the chicken homologue c*SOX*9 also showed male-specific genital ridge expression [55]. Even in the alligator, which has temperature dependent sex determination, *SOX*9 is upregulated in hatchlings incubated at the male, but not the female-determining temperature [59]. *SOX*9 homologues in rainbow trout are expressed predominantly in the adult brain and testis [56]. *SOX*9 therefore has an essential and critical role in the development of the vertebrate testis which need not depend on *SRY*. Thus *SRY* appears to be a mammal-specific switch operating on an highly conserved underlying developmental pathway involving *SOX*9.

However, it is significant that *SOX*9 expression in alligator and chicken begins well after pre-Sertoli cell differentiation and AMH expression, the first signs of testis differentiation [55, 57, 59]. This would suggest that *SOX*9 is not, after all, the first gene in the conserved testis differentiation pathway and may be involved in Sertoli cell organization, rather than determination. If *SRY* has its effect in regulating *SOX*9 activity, there is a real question about what steps precede it in the pathway.

In mammals, female development occurs by default in the absence of the testis determining gene, presumably by activation of genes that control ovary differentiation. The possibility that male determination occurs, not by activation of male-gonadogenesis genes, but suppression of female-gonadogenesis genes, was revisited with the discovery of XY females with an intact *SRY* gene, but with small duplications of the short arm of the X chromosome containing the *DAX1* gene [60–62]. A conserved mouse homologue *Dax1* lies on the mouse X [64]. *DAX1* codes for a member of the nuclear hormone receptor superfamily, and can repress steroid-producing genes [61, 63]. Human *DAX1* is expressed in adrenal glands and gonadal tissue. Mouse *Dax1* is expressed in the adrenals throughout development. *Dax1* is initially expressed in both male and female genital ridges, but rapidly decreases in males to an almost undetectable level by day 12.5 p.c. [64, 65]. This is the same time at which *Sry* expression ceases in the genital ridge, suggesting that *SRY* might work by interfering with *DAX1* function to induce testis development. Perhaps an extra active copy of the *DAX1* gene in human XdupY females is sufficient to repress *SRY* action and prevent male sex determination.

In both species, the expression pattern of *DAX1* is almost identical to that of *SF*1, and this, with the presence of a conserved *SF*1 response element in the *DAX1* promoter region, suggests that the *SF*1 gene directly regulates *DAX*1. However, *Dax*1 expression does not depend on the presence of *Sf*1, since *Dax*1 expression is unaffected in *Sf*1 knock-out mice [66] and also occurs in the absence of *Sf*1 in the fetal rat testis [67]. This would suggest that that *SF*1 and *DAX*1 do not directly interact in the sex-determining pathway, but may act together in activating and regulating steroid hormones during embryogenesis.

A gene has recently been identified on human chromosome 9, in a region deleted in XY sex reversed female patients. The *DMRT1* and *DMRT2* (Doublesex and *MAB-3* related in testis) genes were cloned and characterized and both were found to lie in the smallest deletion interval known to cause gonadal dysgenesis and XY sex reversal [68, 69]. Although no point mutations have been described in sex reversed patients, *DMRT1* is a candidate gene for sex [72]. *DMRT1* shares sequence similarities with genes known to regulate sex in nematodes and insects. *DMRT1* is expressed exclusively in the human testis and is upregulated in the genital ridge at the time of testis determination in mouse [70], suggesting a highly conserved role for this gene in sex determination [71]. Both *DMRT1* and *DMRT2* contain a unique (DM) DNA binding domain, through which the gene is though to act.

DMRT1 may be particularly significant in vertebrate sex determination, as human chromosome 9 bears a large portion of the genes on the Z chromosome which determines sex in chickens [73]. The chicken *DMRT1* homologue maps to the Z chromosome but not the W [reviewed 74], which is compatible with a dosage-sensitive function of this gene in bird sex determination. Chicken *DMRT* is upregulated in the genital ridge during testis determination [70]. The conserved position of *DMRT1* in sexual differentiation from invertebrates to humans would suggest it is an ancient vertebrate sex-determining gene, and holds a critical function in the sex-determining pathway. Further investigations into *DMRT1* in marsupials, monotremes and other animals will help to define the conserved role for this gene in sex determination, which would appear to extend from invertebrates to humans.

Mutations in human *ATRX* cause gonadal dysgenesis and XY sex reversal in affected individuals, in some cases leading to complete male to female sex reversal and female gender assignment [75]. The absence of Müllerian ducts in the *ATRX* mutants indicates that AMH (Anti-Müllerian Hormone) has been expressed, confirming Sertoli cell development within the gonad [43]. The development of the gonad has therefore been interrupted in ATRX mutants at a stage prior to testis organization, but after the testis determining signal from *SRY* and the differentiation of Sertoli and Leydig cells. *ATRX* therefore functions downstream of *SRY* and *AMH* in the male development pathway, perhaps at the level of *SOX*9. Human *ATRX* is located at Xq 13.3 [76] within the conserved region of the X chromosome present on the X in all mammalian species. This suggests that it may have had a role (perhaps dosage-regulated) in gonad development in an ancestral mammal, a hypothesis supported by the finding of a testis-specific copy on the Y chromosome in marsupials [Pask et al., submitted].

Several other sex-reversal syndromes have been observed in human patients, but the genes responsible have not yet been identified. Several autosomal genes also appear to be involved in sex reversal in interspecies mouse backcrosses [77]. The sex determining pathway is obviously much more complex than was first appreciated [78]. Further clues to the role of these and other genes in the pathway may be gained by considering how the sex determining pathway evolved in mammals.

Evolution of the control of the sex determination pathway

Our knowledge of the evolution of mammalian sex chromosomes predicts that genes on the X and Y chromosomes were originally autosomal. Genes such as *SRY*, which are shared by the marsupial Y, must have remained on the Y for at least the 130 MYrBP since marsupials and eutherians diverged, making it likely that they acquired a male-specific function early in mammalian evolution. But genes such as *ZFY*, which are autosomal in marsupials, acquired this role more recently. The origin and relationships of putative sex determining and differentiation genes can be tested by cloning, sequencing and mapping them in marsupials. The roles of SRY and its relatives, and of DAX1, have been reevaluated from an evolutionary standpoint.

The evolution of *SRY*

The evidence suggests that *SRY* is a recent acquisition of the mammalian Y chromomosme, and that it need not act in the same way in all mammals. The differences in sequence, expression and function raise the unpalatable prospect that the mouse *Sry* has evolved a different action from human *SRY*. Marsupial *SRY* may be different again – in fact, there is no direct evidence that *SRY* has a sex determining function in marsupials, and there appears to be no *SRY* in monotremes, or in reptiles and birds.

The male-dominant action of the mammalian testis-determining factor led to the expectation that this gene is a critical and integral part of the sex-determining pathway. However, the absence of a sex-specific SRY from nonmammalian vertebrates implies that SRY has no conserved role in vertebrate sex. Nor is there any evidence for a sex-specific *SRY* in monotremes (A. Pask and P. Western, unpublished observations). At the earliest, then, *SRY* evolved after the monotreme-therian divergence about 170 million years ago. This suggests that *SRY* evolved a control function relatively recently, and is incidental in the sexual development pathway.

Its sex determining function may be even more recent. The discovery of a Y-borne *SRY* in marsupials suggested that this gene is the testis-determining factor in all therian mammals, although testis determination in marsupials does not determine all other sexual dimorphisms as it does in eutherians. While the ubiquitous expression of marsupial *SRY* does not disqualify it from acting as a testis determinant, in the absence of mutation analysis or transgenesis, we cannot be certain that *SRY* is sex determining in marsupials. The sequence of marsupial *SRY* is too poorly conserved outside the HMG box to be sure that the gene serves the same function, and its ubiquitous expression suggests it lacks the regulatory elements present in eutherian *SRY*. Perhaps marsupial *SRY* retains a more primitive form of regulation that has come under tighter control in eutherian mammals, or perhaps it performs some other more general function. It may not, after all, even be sex determining in this group of mammals. Thus *SRY* may have evolved its function a mere 80 million years ago!

Since *SRY* has evolved and been recruited into the sex determining pathway only 170 MYrBP at most, we may be able to learn about its possible function by examining the gene from which it evolved. The *SOX*3 gene was found to be on the X in marsupials, and subsequently in all therian mammals, so must have been on the X in a mammalian common ancestor. Of all its relatives in the SOX gene family, *SRY* shows the most sequence similarity to *SOX*3 within the HMG box – in fact, *SRY* genes from different species are more similar to *SOX*3 than they are to each other. This suggests that *SRY* evolved from *SOX*3 [79]. *SRY* seems to be essentially a truncated *SOX*3. Like other genes on the Y chromosome, it has been mutated and deleted – but has been retained because it found a male-specific function.

*SOX*3 is highly conserved between species, suggesting that it has a critical function in mammalian development [79]. Expression analysis of *SOX*3
in human embryos detected transcripts in the developing brain, spinal cord,
thymus and heart, and *SOX*3 transcripts were detected in several adult
tissues including testis [80]. Mouse *Sox*3 is expressed in the developing
central nervous system and the indifferent gonadal ridge at low levels com-
parable to those of *SRY* [81]. Expression of a chicken homologue c*SOX*3 is
also restricted to the central nervous system. An amphibian homologue,
X*SOX*3 from *Xenopus*, is expressed only in the ovary, and shows highest
expression early in oocyte development [82, 83].

The expression of *SOX*3 in developing gonads in different vertebrates
suggests a conserved role in gonad differentiation in mammals, as well
as a role in the differentiation of central nervous system. However, two
mentally retarded boys with *SOX*3 deletions show testicular development,
excluding *SOX*3 from a role in male sex determination [80]. Perhaps, then,
*SOX*3 is involved in ovary development, as in *Xenopus*, or perhaps it acts
as an inhibitor of testis determination.

It has been postulated that *SRY*, *SOX*3 and *SOX*9 interact to determine
testis [84]. In females, in the absence of *SRY*, *SOX*3 inhibits *SOX*9 and no
testis forms. In males, *SRY* inhibits *SOX*3, permitting *SOX*9 to enact its
testis-determining role (Fig. 3). This hypothesis requires that *SRY* and *SOX*3
are expressed in the same tissue type at the same time. In mouse, *SOX*3 and
SRY are both expressed in the indifferent gonad around day 11.5 p.c., but
there is no information on expression in other mammals. The hypothesis
provides a good explanation of sex reversing mutations in humans, particularly of *SRY*-negative XX males, who may have a mutation in *SOX*3.

Recent evidence supports the idea that SOX9 is negatively regulated, and
that its function is the result of a balance with an inhibitor. An XX sex
reversed male patient has recently been described with a duplication of the
portion of chromosome 17 carrying SOX9 [85]. This finding strongly supports the idea of a double inhibition of SOX9 by SRY, although they do not
reveal the identity of the negative regulator – is it SOX3?

The hypothesis that *SOX*3 is involved in the sex-determination circuitry
lends itself to a sensible account of the evolution of the male-dominant
testis determination by *SRY* from an earlier dosage-dependent system. It
has been proposed that in an ancestral mammal homozygotes for wild
type *SOX*3 were female, whereas heterozygotes for a null allele were
male; the 2:1 dosage difference determined sex via a differential effect on
*SOX*9 activity. In *Xenopus*, *SOX*3 may act to inhibit *SOX*9 and permit ovary
determination. Since *SOX*3 is subject to X inactivation in eutherian mammals, this dosage-sensitive system must have been supplanted, but perhaps
in marsupials, which show incomplete X inactivation, *SOX*3 dosage could
still determine sex. Further investigations into *SOX*3 function and expression in other mammals will shed light on its role in sex determination and
its interaction with *SRY* and *SOX*9.

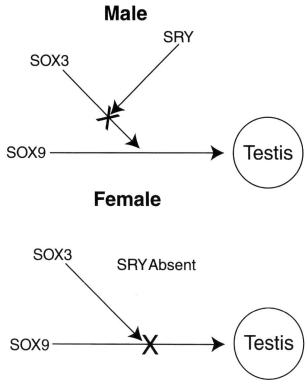

Figure 3. Proposed mechanism of *SOX3*, *SOX9* and *SRY* interactions in sex determination. In the presence of *SRY*, *SOX3* is prevented from repressing *SOX9*, which in turn results in the development of a testis. In the absence of *SRY*, *SOX3* can inhibit *SOX9*, which in turn prevents the formation of a testis, leading to female development.

Dosage-sensitive action of *DAX*1

Gene-dosage differences seem to be a recurring theme in sex determination systems in a wide variety of animal species, and are well characterised in *Drosophila* and *Caenorhabditis elegans*. It was proposed that X chromosome inactivation evolved in mammals as a sex determination, rather than a dosage compensation system [86]. The effects of X chromosome dosage on the sexual phenotype in XO and XXY marsupials also suggested that a dosage-sensitive gene is involved in sexual differentiation at least in this mammal group.

This intriguing idea was put to one side in the excitement of cloning the testis-determining factor, but the identification of the dosage-sensitive sex-reversing gene *DAX*1 on the X led to speculation that this gene might regulate ovarian development in mammals by virtue of its dosage on the X in males and females. Two copies of this gene could determine female development, and one male development. This could be the case in marsupials, in which X inactivation is incomplete; however, *DAX1* must be subject to inactivation in humans and mice, since XXY are male in both species, so the dosage of *DAX1* is equal in males and females.

It is still possible that *DAX1* represented an ancestral dosage-sensitive gene that functioned in sex determination in a mammalian ancestor before this locus was recruited into the X inactivation system. If *DAX1* were involved in an ancestral mammalian DSS mechanism, it should lie on the conserved region of the X chromosome present in all mammalian groups. *DAX1* is also a good candidate for the X-linked dosage sensitive sex differentiation locus found on the marsupial X chromosome, which is responsible for the intersexual phenotypes of XO and XXY marsupials.

Both hypotheses predict that *DAX1* should lie on the marsupial X chromosome. A marsupial *DAX1* homologue was isolated and shown to share 73% amino acid identity to human *DAX1* within the ligand-binding domain. This gene was mapped by fluorescence *in situ* hybridization to chromosome 5 in the tammar wallaby, colocalizing with several other genes on the short arm of the human X [87]. This implies that *DAX1* was not present on the X chromosome in the mammalian common ancestor, but is part of a recent addition to the eutherian X. An autosomal *DAX1* gene could never have functioned in a dosage-sensitive role in marsupials, and cannot be the X-linked dosage-sensitive sex differentiation gene on the marsupial X chromosome.

This does not necessarily exclude *DAX1* from a dosage-sensitive role in eutherian sex determination, but it would mean that this role must have been confined to a time after the marsupial-eutherian divergence 130 MyrBP, and the recruitment of *DAX1* into the X inactivation system. It is difficult to see why a gene recruited to the X chromosome would take on a dosage-sensitive sex-determining function, which has the ability to override the male determining signal from the *SRY* gene. It seems more likely that *DAX1* has a role in gonad determination that is independent of its position. The level of *DAX1* expression in human gonads must be critical, as two active copies can override the male-determining signal from *SRY*. It is therefore likely that its recruitment to the X inactivation system was accompanied by upregulation, as has been described for Smcx [88]. Comparing the activity of DAX1 in marsupial gonads, in which there are two active (autosomal) loci, may help to explain how this gene became dosage sensitive.

ATRX – the ancestral mammalian sex-determining locus?

SRY is a recent addition to the sex determining pathway being restricted to the marsupial and eutherian lineages. As yet, no *SRY* homologue has been detected in monotremes. Perhaps dosage of *SOX3* preceded *SRY* as the testis determining factor – and perhaps another gene preceded this system.

We have recently identified a Y-borne homologue of the sex-reversing gene *ATRX* which appears to be a good candidate for the sex determining gene in marsupials, and may represent an ancestral mammalian sex determining gene.

Mutations within the *ATRX* gene cause numerous defects including severe psychomotor retardation, characteristic facial features, α-thalassemia, genital abnormalities, and several other congenital irregularities [89]. The level of gonadal dysgenesis in affected individuals is variable, in some cases leading to complete male to female sex reversal and female gender assignment [75] confirming a role for *ATRX* in testis development. Homologues of *ATRX* were isolated from marsupials, characterized and mapped. Surprisingly, homologues of the *ATRX* gene were found to be located on both the X (*ATRX*) and Y (*ATRY*) chromosomes in marsupials. Expression analysis of *ATRX* and *ATRY* in the wallaby showed distinct, yet complimentary patterns of expression. *ATRX* was expressed in every tissue sampled from developing and adult wallabies, but was absent from the developing gonad. Conversely, *ATRY* expression was present only in the developing and adult testis and was absent from all other tissues (Pask et al., submitted). This would suggest that marsupial ATRX has maintained a general developmental role, while ATRY has a specialized function in the development of the testis. Here, marsupials provide us with a unique opportunity to study the control over genes in the sex determining pathway since *ATRY* would contain a testis-specific promoter region. Furthermore, ATRY showing testis-specific expression and being located on the marsupial Y chromosome represents a better candidate testis determining gene for this group of mammals than the ubiquitously expressed *SRY* gene. Of particular interest is the location of *ATRY* in monotremes which lack the *SRY* gene. If *ATRY* is present on the Y in monotremes it would represent a testis determining candidate in this lineage.

The location of *ATRX* gene homologues on the X and Y in marsupials would suggest that *ATRX* and *ATRY* were present on the proto X and Y chromosome in the common ancestor to both groups. In addition, the expression of *ATRY* in the developing and adult marsupial testis would indicate that homologues in the common ancestor maintained a role in testis development. This suggests that ATRY had an ancestral role in testis development, possibly before the evolution of SRY.

DMRT – the ancestral vertebrate sex-determining locus?

DMRT1 shows sexually dimorphic expression in embryonic gonads of mice, birds and alligators, being upregulated in the male gonads of all three species. This would indicate DMRT1 to have a highly conserved role in vertebrate testis determination [70]. *DMRT1*, therefore, represents the most ancient sex-determining gene known to date. The location of *DMRT1* on

the bird Z chromosome raises the possibility that it acts as a sex determining switch gene in this species – and possibly in ZZ/ZW reptiles too.
The *SRY* gene homologue *SOX3*, and *ATRY*, too, are likely to be on chick
chromosome 1 or 4 along with other mammalian X-linked genes. Both sex
determining mechanisms (XX/XY-ZZ/ZW) have evolved independently
but still appear to trigger a common sexual differentiation pathway that
involves *SOX9* and *DMRT* homologues. The two sex determining strategies
XX/XY vs. ZZ/ZW have evolved around these two different loci, that both
trigger a conserved gonadogenesis pathway in two completely different
manners.

Conclusion

Although testis determination seems to be very similar at the histological
level in all vertebrates, the control of the sex-determining pathway has
evidently changed radically in evolution. Comparisons of the location,
sequence and expression of sex-determining genes in other vertebrates, and
particularly between the three extant mammalian groups, has provided
many insights into their importance and role in the vertebrate sex-determining pathway, as well as informing us of possible evolutionary changes in
this control.

Comparative mapping has allowed us to assess the conserved role of
gene in the sex determining pathway between distantly related groups.
Comparisons of the candidate mammalian sex-determining genes revealed
that *ZFY*, the first candidate TDF, is a recent addition to the eutherian sex
chromosomes and therefore could not be the universal mammalian testis-
determining factor. In contrast, *SRY* was found to lie on the Y chromosome
in marsupials as well as eutherians.

There is no evidence for a sex specific *SRY* gene outside of Class
Mammalia, indicating that this gene has only recently acquired its male-
determining switch role in therian mammals, while other vertebrates use
different mechanisms for triggering the male sexual development pathway.

At what time in mammalian evolution SRY acquired its male determining role is undetermined. As yet monotreme mammals have not been
demonstrated to posses a sex specific *SRY* gene, suggesting that *SRY* evolved its male determining function after the divergence of monotremes and
therian mammals about 170 million years ago. However this figure could
be much more recent as in marsupials it is yet to be demonstrated that *SRY*
is sex determining, and it is still possible that in this mammal group, testis
determination is triggered by some other gene.

It is clear, that *SRY* appears to be no more than a trigger of the male development pathway, and its action may be very indirect. The poor sequence
conservation of SRY between species both within and flanking the HMG
box, suggests that *SRY* may encode a repressor rather than a transcriptional

activator. This is in direct contrast to the related SOX9 gene that is highly conserved, and includes a transactivating domain suggesting its action is much more direct in sex determination. An inadequate dose of *SOX9* causes male to female sex reversal, while overproduction causes XX female to male gonadal sex reversal. The male-specific transcriptional upregulation of *SOX9* in the gonad of all mammals, reptiles and birds suggest it to have an important role in testis organization, if not testis determination.

Little attention has been paid to the potential role of *SOX3* in sex determination, despite its close evolutionary relationship with *SRY* and *SOX9*. Furthermore its expression in developing gonad and ability to bind to the same target site as SRY raises the question of its interaction with SRY in sex determination. *SOX3* is highly conserved among eutherian and marsupial mammals, and is expressed in the testis in humans and mouse. However, bird *SOX3* is not expressed in the developing gonad, and in amphibians expression is detected in the developing ovary, and appears to be involved in oocyte development. This gene may have originally had a mammalian dosage sensitive role in female determination, which was taken over in therian mammals when *SRY* was differentiated on the Y chromosome.

DAX1 is a candidate for an ancient dosage-sensitive sex determiner. However, the autosomal location of *DAX1* in marsupials suggests that it was not present on the ancestral sex chromosome but is rather a recent addition to the eutherian X. Why and how *DAX1* acquired this dosage sensitive function is still unknown, as is the mechanism by which it triggers female development. Analysis of the autosomal *DAX1* expression in marsupials may help to define its role in mammalian sexual development.

The location of an *ATRX* homologue on the marsupial Y chromosome is particularly interesting as it has a role in human testis development. Marsupial *ATRY* has a specific role in testis development compared to its X-linked homologue, which maintains roles in several tissues. Therefore marsupial *ATRY* provides us with a unique opportunity in which to investigate the specific promoter elements and regions of the gene essential for its role in the testis development pathway, partitioned from its wider role. Of particular interest is the location of *ATRX* homologues in the monotemes, which lack an *SRY* gene. It is possible that in ancestral mammals sex was determined by *ATRY* and not *SRY*.

DMRT1 represents the most ancestral sex-determining gene cloned to date. The location of *DMRT1* on the bird Z chromosome raises the possibility that this gene determines sex in birds by virtue of its dosage. If this were the case, such a pivotal gene in vertebrate gonad development would be expected to be under the direct control of SRY in mammals; however, this is currently untested.

As it currently stands we still lacking fundamental knowledge about the role of each gene in sex determination. The action of SRY is dependent on regulatory sequences and could also be affected by interactions with *SIP1* and calmodulin. The coexpression of *SOX3* and *SRY* in the gonadal ridge would suggest that the two genes compete for the same binding site. It is suggested that *SRY* binding would block *SOX3* from initiating female development, and could possibly lead to the activation of *ATRX* and *SOX9*, essential for the organisation of the testis. *SF1* is responsible for activation of *AMH* expression from the pre-Sertoli cells, leading to Müllerian duct regression and the masculinization of the gonads. In females the story is more vague with *DAX1* being the only gene identified with a role in ovary determination. *SOX3*, as indicated above, could also play a role in female development. At what level each of the interactions described above occur is still undetermined (Fig. 4). The most significant event in sex determination is the development of the Sertoli cells in the vertebrate testis. As yet the gene responsible for this pivotal role is unidentified. The sexually dimorphic expression of *DMRT1* in mammals, birds and reptiles highlights the long-term association of this gene with the sex-determining pathway, and raises the possibility that this gene determines Sertoli cell differentiation in all vertebrates.

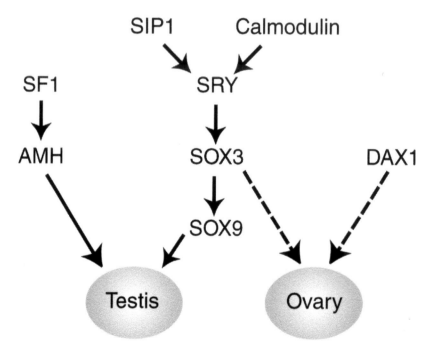

Figure 4. Proposed interactions between the sex-determining genes. The exact nature of each interaction is still unknown. Solid arrows indicate the male development pathway. Dashed arrows indicate the female development pathway.

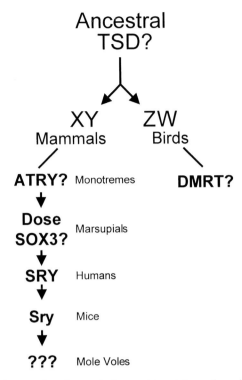

Figure 5. Comparative mapping of genes in the sex determining pathway has uncovered a conserved set of genes that have occupied roles in gonadal development through millions of years of evolution. The control over this fundamental pathway, however, varies enormously in the mechanisms and genes involved even between closely related species.

Comparative studies of the genes involved in sex determination have therefore allowed us to determine that some genes, such as *SOX9*, have critical conserved functions in sex determination, whereas others, such as *SRY*, have recently evolved a role in the sex-determining pathway of one vertebrate lineage. Obviously the pathway is more complex and more circuitous than was originally envisaged, and there are many genes and more question marks (Fig. 4). It would appear from comparative studies that there has been rapid evolution of the control of a highly conserved vertebrate sex determining pathway (Fig. 5). Further comparative studies may assist our understanding of the role of genes associated with the pathway, and may ultimately help us to construct the entire mammalian sex-determining pathway and identify genes which have functions in common with other pathways of vertebrate organogenesis.

References

1 Hope RM, Cooper S and Wainwright B (1990) Globin macromolecular sequences in marsupials and monotremes. In: Mammals from Pouches and Eggs: Genetic Breeding and Evolution of Marsupials and Monotremes, Graves JAM, Hope RM and Cooper DW (eds), CSIRO Australia, Melbourne

2 Sharman GB, Hughes RL and Cooper DW (1990) The chromosomal basis of sex differentiation in marsupials. In: Mammals from Pouches and Eggs: Genetics, Breeding and Evolution of Marsupials and Monotremes, pp. 309–324, Graves JAM, Hope RM and Cooper DW (eds), CSIRO Australia, Melbourne

3 Cooper DW, Johnston PG, Watson JM and Graves JAM (1993) X-inactivation in marsupials and monotremes. *Dev Biol* 4: 117–128

4 Vogt PH (1997) Report of the third international workshop on Y chromosome mapping. *Cytogenet Cell Genet* 79: 1–20

5 Graves JAM, Wakefield MJ and Toder R (1998) The origin and evolution of the pseudo-autosomal regions of human sex chromosomes. *Hum Mol Genet* 7: 1991–1996

6 Sharp P (1982) Sex chromosome pairing during male meiosis in marsupials. *Chromosoma* 86: 27–47

7 Murtagh CE (1977) A unique cytogenetic system in monotremes. *Chromosoma* 65: 37–57

8 Graves JAM (1995) The origin and function of the mammalian Y chromosome and Y-borne genes – an evolving understanding. *Bioessays* 17: 311–320

9 Foster JW, Brennan FE, Hampikian GK, Goodfellow PN, Sinclair AH, Lovell-Badge R et al (1992) Evolution of sex determination and the Y chromosome: *SRY*-related sequences in marsupials. *Nature* 359: 531–533

10 Delbridge ML, Ma K, Subbarao MN, Cooke HJ, Bhasin S and Graves JAM (1998) Evolution of mammalian HNRPG and its relationship with the putative azoospermia factor RBM. *Mamm Genome* 9: 168–170

11 Mitchell MJ, Woods DR, Wilcox SA, Graves JAM and Bishop CE (1991) Marsupial Y chromosome encodes a homologue of the mouse Y-linked candidate spermatogenesis gene Ube1y. *Nature* 359: 528–531

12 Agulnik AI, Mitchell MJ, Lerner JL, Woods DR and Bishop CE (1994) A mouse Y chromosome gene encoded by a region essential for spermatogenesis and expression of male-specific minor histocompatibility antigens. *Hum Mol Genet* 3: 873–878

13 Toder R and Graves JAM (1998) CSF2RA, ANT3 and STS are autosomal in marsupials – implications for the origin of the pseudoautosomal region of mammalian sex chromosomes. *Mamm Genome* 9: 373–376

14 Ohno S (1967) Sex chromosomes and sex linked genes. Springer, New York

15 Graves JAM (1998) Evolution of the mammalian Y chromosome and sex determining genes. *J Exp Zool* 281: 472–481

16 Charlesworth B (1991) The evolution of sex chromosomes. *Science* 251: 1030–1033

17 Graves JAM, Disteche CM and Toder R (1998) Gene dosage in the evolution and function of mammalian sex chromosomes. *Cytogenet Cell Genet* 80: 94–103

18 Page DC, Mosher R, Simpson EM, Fisher EM, Mardon G, Pollack J et al (1987) The sex-determining region of the human Y chromosome encodes a finger protein. *Cell* 51: 1091–1104

19 Sinclair AH, Foster JW, Spender JA, Page DC, Palmer M, Goodfellow PN et al (1988) Sequences homologous to *ZFY*, a candidate human sex-determining gene, are autosomal in marsupials. *Nature* 336: 780–783

20 Koopman P, Gubbay J, Collignon J and Lovell-Badge R (1989) *ZFY* gene expression patterns are not compatible with a primary role in mouse sex determination. *Nature* 342: 940–942

21 Palmer MS, Sinclair AH, Berta P, Ellis NA, Goodfellow PN, Abbas NE et al (1989) Genetic evidence that *ZFY* is not the testis-determining factor. *Nature* 342: 937–939

22 Sinclair AH, Berta P, Palmer MS, Hawkins JR, Griffiths BL, Smith MJ et al (1990) A gene from the human sex determining region encodes a protein with homology to a conserved DNA binding motif. *Nature* 346: 240–244

23 Hawkins JR (1994) Sex determination. *Hum Mol Genet* 3: 1463–1467

24 Gubbay J, Collignon J, Koopman P, Capel B, Economou A, Munsterberg A et al (1990) A gene mapping to the sex-determining region of the mouse Y chromosome is a member of a novel family of embryonically expressed genes. *Nature* 346: 245–250

25 Koopman P, Gubbay J, Vivian N, Goodfellow P and Lovell-Badge R (1991) Male development of chromosomal female mice transgenic for *SRY*. *Nature* 351: 117–121

26 Foster JW, Brennen FE, Hampikian GK, Goodfellow PN, Sinclair AH, Lovell-Badge R et al (1992) The human sex determining gene *SRY* detects homologous sequences on the marsupial Y chromosome. *Nature* 359: 531–533

27 Foster JW and Graves JAM (1994) An SRY related sequence of the Marsupial X chromosome: Implications for the evolution of the mammalian testis-determining gene. *Proc Natl Acad Sci USA* 91: 1927–1931

28 O'Neill RJ, Brennan FE, Delbridge ML, Crozier RH and Graves JAM (1998) *De novo* insertion of an intron into the mammalian sex determining gene, *SRY*. *Proc Natl Acad Sci USA* 95: 1653–1657

29 Dubin RA and Ostrer H (1994) *SRY* is a transcriptional activator. *Mol Endocrinol* 8: 1182–1192

30 Bowles J, Cooper L, Berkman J and Koopman P (1999) Sry requires a CAG repeat domain for male sex determination in Mus musculus. *Nature Genetics* 22: 405–408

31 Poulat F, Desantabarbara P, Desclozeaux M, Soullier S, Moniot B, Bonneaud N et al (1997) The human testis determining factor *SRY* binds a nuclear factor containing PDZ protein interaction domains. *J Biol Chem* 272: 7167–7172

32 Koopman P, Munsterberg A, Capel B, Vivian N and Lovell-Badge R (1990) Expression of a candidate sex-determining gene during mouse testis differentiation. *Nature* 348: 450–452

33 Capel B, Swain A, Nicolis S, Hacker A, Walter M, Koopman P et al (1993) Circular transcripts of the testis-determining gene *Sry* in adult mouse testis. *Cell* 73: 1019–1030

34 McElreavy K, Vilain E, Abbas N, Costa JM, Souleyreau N, Kucheria K et al (1992) XY sex reversal associated with a deletion 5′ to the SRY "HMG box" in the testis-determining region. *Proc Natl Acad Sci USA* 89: 11016–11020

35 Harry JL, Koopman P, Brennan FE, Graves JAM and Renfree MB (1995) Wide spread expression of the testis determining gene *SRY* in a marsupial. *Nat Genet* 11: 347–349

36 Ferrari S, Harley V, Pontiggia A, Goodfellow P, Lovell-Badge R and Bianchi ME (1992) A sharp angle in DNA is the major determinant in DNA recognition by the *SRY* protein as it is for HMG 1 protein. *EMBO J* 11: 4497–4509

37 Pontiggia A, Whitfield S, Goodfellow PN, Lovellbadge R and Bianchi MR (1995) Evolutionary conservation in the DNA-binding and bending properties of HMG boxes from *SRY* proteins of primates. *Gene* 154: 277–280

38 Cohen DR, Sinclair AH and McGovern JD (1994) The SRY protein enhances transcription of a FOS-related antigen 1 promoter construct. *Proc Natl Acad Sci USA* 91: 4372–4376

39 Koopman P (1995) The molecular biology of *SRY* and its role in sex determination in mammals. *Reprod Fertil Dev* 7: 713–722

40 McElreavey K, Rappaport R, Vilain E, Abbas N, Richaud F, Lortat-Jacob S et al (1992) A minority of 46,XX true hermaphrodites are positive for the Y-DNA sequence including *SRY*. *Hum Genet* 90: 121–125

41 Just W, Rau W, Vogel W, Akhverdian M, Fredga K, Graves JAM et al (1995) Absence of *SRY* in a species of the vole *ellobius*. *Nature Genet* 11: 117–118

42 Harley VR, Lovell-Badge R, Goodfellow PN and Hextall PJ (1996) The HMG box of *SRY* is a calmodulin binding domain. *FEBS Lett.* 391: 24–28

43 Lee MM and Donahoe PK (1993) Mullerian inhibiting substance: A gonadal hormone with multiple functions. *Endocr Rev* 14: 152–160

44 Haqq CM, King CY, Ukiyama E, Falsafi S, Haqq TN, Donahoe PK et al (1994) Molecular basis of mammalian sex determination: activation of Mullerian inhibitory substance gene expression by *SRY*. *Science* 266: 1494–1500

45 Jost A, Vigier B, Prepin J and Perchellet J (1972) Freemartins in cattle: the first steps of sexual organogenesis. *J Reprod Fertil* 29: 349–379

46 Ingraham HA, Lala DS, Ikeda Y, Luo XR, Shen WH, Nachtigal MW et al (1994) The nuclear receptor steroidagenic factor 1 acts at multiple levels of the reproductive axis. *Genes Dev* 8: 2302–2312

94

A. Pask and J.A. Marshall Graves

47 Shen WH, Moore CCD, Ikeda Y, Parker KL and Ingraham H (1994) Nuclear receptor steroidagenic factor 1 regulates the Mullerian inhibiting substance gene: a link to the sex determination cascade. *Cell* 77: 651–661
48 Tommerup N, Schempp W, Meinecke P, Pedersen S, Bolund L, Brandt C et al (1993) Assignment of an autosomal sex reversal locus (SRA1) and campomelic dysplasia (CMPD1)
to 17q24.3-q25.1. *Nature Genet* 4: 170–174
49 Foster JW, Dominquez MA, Guioli S, Kwok C, Weller PA, Stevanovic M et al (1994)
Campomelic dysplasia and autosomal sex reversal caused by mutations in an *SRY*-related
gene. *Nature* 372: 525–530
50 Wagner T, Wirth J, Meyer J, Zabel B, Held M, Zimmer J et al (1994) Autosomal sex reversal and campomelic dysplasia are caused by mutations in and around the *SRY*-related gene
*SOX*9. *Cell* 79: 1111–1120
51 Wright E, Hargrave R, Christiansen J, Cooper L, Kun J, Evans T et al (1995) The *SRY*-
related *Sox*9 gene is expressed during chondrogenesis in mouse embryos. *Nature Genet* 9:
15–20
52 Bell DM, Leung KKH, Wheatley SC, Ng LJ, Zhou S, Ling KW et al (1997) *SOX*9 directly
regulates the type-II collagen gene. *Nature Genet* 16: 174–178
53 Lefebre V, Huang W, Harley V, Goodfellow PN and de Crombrugghe B (1997) *SOX*9 is a
potent activator of the chondrocyte-specific enhancer of the Proa1(II) collagen gene. *Mol
Cell Biol* 17: 2336–2346
54 Sudbeck P, Schmitz ML, Baeuerle PA and Scherer G (1996) Sex reversal by loss of the
C-terminal transactivation domain of human *SOX*9. *Nature Genet* 13: 230–232
55 Kent J, Wheatley SC, Andrews JE, Sinclair AH and Koopman P (1996) A male-specific role
for *SOX*9 in vertebrate sex determination. *Development* 122: 2813–2822
56 Takamatsu N, Kanda H, Ito M, Yamashita A, Yamashita S and Shiba T (1997) Rainbow trout
*SOX*9 cDNA cloning, gene structure and expression. *Gene* 202: 167–170
57 Oreal E, Pieau C, Mattei M, Josso N, Picard J, Carre-eusebe D et al (1998) Early expression
of AMH in chicken embryonic gonads precedes testicular *SOX*9 expression. *Dev Dynam*
21: 522–532
58 Morais da Silva S, Hacker A, Harley V, Martineau J, Capel B, Goodfellow P et al (1996)
*SOX*9 expression during gonadal development implies a conserved role for the gene in
Sertoli cell differentiation in mammals and birds. *Nature Genet* 14: 62–68
59 Western PS, Harry JL, Graves JAM, Sinclair AH (1999) Temperature-dependent sex determination: Upregulation of SOX9 expression after commitment to male development. *Dev
Dynamics* 214: 171–177
60 Bardoni B, Zanaria E, Guioli S, Floridia G, Worley K, Tonini G et al (1994) A dosage sensitive locus at chromosome Xp21 is involved in male to female sex reversal. *Nature Genet*
7: 497–501
61 Zanaria E, Muscatelli F, Bardoni B, Strom T, Guioli S, Guo W et al (1994) An unusual
member of the nuclear hormone receptor super family responsible for X-linked adrenal
hypoplasia congenita. *Nature* 372: 635–641
62 Muscatelli F, Strom T, Walker AP, Zanaria E, Recan D, Meindi A et al (1994) Mutations
in the *DAX*-1 gene give rise to both X-linked adrenal hypoplasia congenita and hypogonadatrophic hypogonadism. *Nature* 372: 672–676
63 Zazopoulos E, Lalli E, Stocco DM and Sassone-Corsi P (1997) DNA binding and transcriptional repression by *DAX*-1 blocks steroidogenesis. *Nature* 390: 311–315
64 Swain A, Zanaria E, Hacker A, Lovellbadge R and Camerino G (1996) Mouse *DAX*1 expression is consistent with a role in sex determination as well as in adrenal and hypothalamus
function. *Nature Genet* 12: 404–409
65 Guo WW, Burris TP and McCabe ERB (1995) Expression of *DAX*-1, the gene responsible
for X-linked adrenal hypoplasia congenita and hypogonadatrophic hypogonadism, in the
hypothalamic-pituary-adrenal gonadal axis. *Biochem Mol Med* 56: 8–13
66 Ikeda Y, Swain A, Weber TJ, Hentges KE, Zanaria E, Lalli E et al (1996) Steroidogenic
factor 1 and *DAX*-1 co-localise in multiple cell lineages – potential links in endocrine development. *Mol Endocrinol* 10: 1261–1272
67 Majdic G and Saunders PTK (1996) Differential patterns of expression of *DAX*-1
and steroidogenic factor 1 (*SF*1) in the foetal rat testis. *Endocrinology* 137: 3586–
3589

68 Veitia R, Nunes M, Brauner R, Doco-Fenzy M, Joanny-Flinois O, Jaubert F, Lortat-Jacob S, Fellous M and McElreavey (1997) Deletions of distal 9p associated with 46,XY male to female sex reversal: definition of the breakpoints at 9p23.3-p24.1. *Genomics* 41: 271–274
69 McDonald MT, Flejter W, Sheldon S, Putzi MJ and Gorski JL (1997) XY sex reversal and gonadal dysgenesis due to 9p24 monosomy. *Am J Med Genet* 73: 321–326
70 Smith CA, McClive PJ, Western PS, Reed KJ and Sinclair AH (1999) Conservation of a sex determining gene. *Nature* 402: 601–602
71 Raymond CS, Shamu CE, Shen MM, Seifert KJ, Hirsch B, Hodgkin J and Zarkower D (1998) Evidence for evolutionary conservation of sex-determining genes. *Nature* 391: 691–695
72 Raymond CS, Parker ED, Kettlewell JR, Brown LG, Page DC, Kusz K, Jaruzelska J, Reinberg Y, Flejterg WL, Bardwell VJ, Hirsch B and Zarkower D (1999) A region of human chromosome 9p required for testis development contains two genes related to known sexual regulators. *Hum Mol Genet* 8: 989–996
73 Nanda I, Sick C, Munster U, Kaspers B, Schartl M, Staeheli P et al (1998) Sex chromosome linkage of chicken and tuck type I interferon genes: further evidence of evolutionary conservation of the Z chromosome in birds. *Chromosoma* 107: 204–210
74 Nanda I, Shan Z, Schartl M, Burt DW, Koehler M, Nothwang H-G et al (1999) 300 million years of conserved synteny between chicken Z and human chromosome 9. *Nature Genet* 21: 258–259
75 Ion A, Telvi L, Chaussain JL, Galacteros F, Valayer J, Fellous M et al (1996) A novel mutation in the putative DNA helicase XH2 is responsible for male-to-female sex reversal associated with an atypical form of the ATRX syndrome. *Am J Hum Genet* 58: 1185–1191
76 Villard L, Gecz J, Colleaux L, Lossi AM, Chelly J, Ishikawa-Brush Y et al (1995) Construction of a YAC contig spanning the Xq13.3 subband. *Genomics* 26: 115–122
77 Eicher EM, Washburn LL, Schork NJ, Lee K, Shown EP, Xu X et al (1996) Sex-determining genes on mouse autosomes identified by linkage analysis of C57BL/6J-YPOS sex reversal. *Nature Genet* 14: 206–209
78 Goodfellow P (1983) Sex is simple. *Nature* 304: 221
79 Foster JW and Graves JAM (1994) An SRY-related sequence on the marsupial X chromosome: Implications for the evolution of the mammalian testis-determining gene. *Proc Natl Acad Sci USA* 91: 1927–1931
80 Stenovic M, Lovell-Badge R, Collignon J and Goodfellow PN (1993) *SOX*3 is an X-linked gene related to *SRY*. *Hum Mol Genet* 2: 2013–2018
81 Collignon J, Sockanathan S, Hacker A, Cohen-Tannoudji M, Norris D, Rastan S et al (1996) A comparison of the properties of *SOX*3 with *SRY* and two related genes, *SOX*1 and *SOX*2. *Development* 122: 509–520
82 Koyano S, Ito M, Takamatsu N, Takiguchi S and Shiba T (1997) The *Xenopus SOX*3 gene expressed in oocytes of early stages. *Gene* 188: 101–107
83 Penzel R, Oschwald R, Chen YL, Tacke L and Grunz H (1997) Characterisation and early embryonic expression of a neural specific transcription factor X*SOX*3 in *Xenopus laevis*. *Int J Dev Biol* 41: 667–677
84 Graves JAM (1998) Interactions between *SRY* and *SOX* genes in mammalian sex determination. *Bioessays* 20: 264–269
85 Huang B, Wang S, Lamb AN, Bartley J (1999) Autosomal XX sex reversal caused by duplication of SOX9. *J Med Genet* 87: 349–353
86 Chandra HS (1984) A model for mammalian male determination based on a passive Y chromosome. *Mol Gen Genet* 193: 384–388
87 Pask A, Toder R, Wilcox SA, Camerino G and Graves JAM (1997) The candidate sex reversing *DAX*-1 gene is autosomal in marsupials – implications for the evolution of sex determination in mammals. *Genomics* 41: 422–426
88 Lingenfelter PA, Adler DA, Poslinski D, Thomas S, Elliott RW, Chapman VM and Disteche CM (1998) Escape from X inactivation of *Smcx* is preceded by silencing during mouse development. *Nature* 18: 212–213
89 Gibbons RJ, Picketts DJ, Villard L and Higgs DR (1995) Mutations in a putative global transcriptional regulators cause X-linked mental retardation with – thalassemia (ATR-X syndrome). *Cell* 80: 837–845

Genes and Mechanisms in Vertebrate Sex Determination
ed. by G. Scherer and M. Schmid
© 2001 Birkhäuser Verlag Basel/Switzerland

An overview of factors influencing sex determination and gonadal development in birds

Michael Clinton[1] and Lynne C. Haines[2]

[1] *Department of Gene Expression & Development, Roslin Institute, Roslin, Midlothian EH25 9PS, UK*
[2] *Comparative and Developmental Genetics Section, MRC, Human Genetics Unit, Western General Hospital, Edinburgh EH4 2XU, UK*

Summary. The morphological development of the embryonic gonads is very similar in birds and mammals, and recent evidence suggests that the genes involved in this process are conserved between these classes of vertebrates. The genetic mechanism by which sex is determined in birds remains to be elucidated, although recent studies have reinforced the contention that steroids may play an important role in the structural development of the testes and ovaries in birds. So far, few genes have been assigned to the avian sex chromosomes, but it is known that the Z and W chromosomes do not share significant homology with the mammalian X and Y chromosomes. The commercial importance of poultry breeding has motivated considerable investment in developing physical and genetic maps of the chicken genome. These efforts, in combination with modern molecular approaches to analyzing gene expression, should help to elucidate the sex-determining mechanism in birds in the near future.

Introduction

For many years the accessibility and resilience of the chick embryo made this system the preferred model for studies on vertebrate development. However, since the advent of modern molecular techniques, the paucity of genetic resources and the difficulties in adapting transgenic technologies to birds has seen the mouse supersede the chick as the model system of choice. This legacy has meant that the morphological development and endocrine control of secondary sexual differentiation is as well understood in birds as in mammals. Unfortunately, it has also meant that the understanding of the genetic control of sex determination and gonadal differentiation in birds lags far behind that in mammals. In recent years there has been renewed interest in investigating these aspects of avian development, and while the primary motivation has been commercial, the impetus has derived from advances in the mammalian field, particularly the identification of the testis-determining gene. This article will review recent progress in sex determination and gonadal development in birds.

General gonadal development

There are many shared features between gonadogenesis in birds and in mammals. In both, the gonads arise from a ridge of tissue, known as the genital ridge, which appears on the surface of the developing mesonephros [1]. The cells which contribute to the developing gonad are derived from the mesenchymal blastema of the genital ridge, from the overlying coelomic epithelium and from the mesonephros; however, the exact contribution from each is under debate [2–4]. During the initial stages of gonadal development there are no obvious differences between males and females, and this is designated the indifferent period. At a particular point in gonadogenesis, development in the male and female diverges: the male gonad displays testicular characteristics, whereas the female gonad displays ovarian characteristics. While the fully developed organs appear dramatically different in structure, the testis and ovary are essentially similar in function and at a cellular level. In both males and females the cells of the indifferent gonad differentiate into the specialized supporting and steroidogenic cells which form tissue cords around the germ cells [2]. It is the localization of these cords within the gonads that leads to the distinctive structural differences between the ovary and testis. In generalized terms, the chick gonads are considered to be composed of two major components, an inner medulla and an outer cortex [2, 5, 6]. In the testis, the cords of cells form in the medullary region, which continues to develop, and the cortex regresses, whereas in the ovary the early medullary cords degenerate and secondary cords develop in an expanded cortex (Fig. 1). At a molecular level, this process can be considered to be under the control of two distinct genetic programs: (i) an underlying gene cascade which regulates the differentiation of the specialized cells of the testis and ovary, and (ii) super-

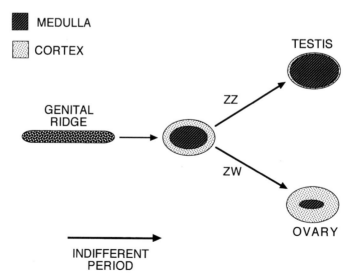

Figure 1. Relative contribution of the medullary and cortical components to the developing gonads.

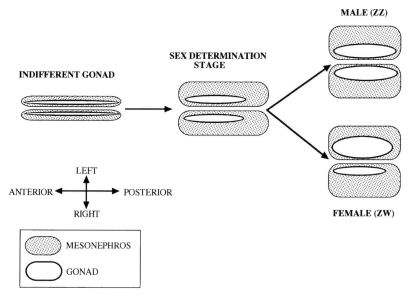

Figure 2. Schematic illustrating the development of the male and female avian gonads. Prior to gonad differentiation male and female embryos are indistinguishable. After gonad differentiation individual embryos can be sexed on the basis of the asymmetrical development of the female gonads.

imposed on this a sex-determining mechanism which controls the decision to follow the pathway of testicular or ovarian development. As previously noted, the development of individual gonads is similar in birds and mammals at a morphological level. However, a major difference in gonadogenesis between these classes is that only the left ovary fully develops in most birds (Fig. 2). The right ovary differentiates to the stage where it produces a female pattern of hormones and then ceases to develop further. Instances where the left ovary is removed or damaged result in the development of the right gonad as a testis or ovo-testis [7, 8], and it has been suggested that the right female gonad only has the capacity to develop as a testis. However, if the left ovary is removed early in embryonic development, the right gonad will develop as a normal ovary, demonstrating an inherent capacity to form an ovary, if only for a restricted period [9, 10].

The right-left asymmetry seen in ovarian development also extends to the distribution of the germ cells, with approximately 70% being found in the left gonad in both sexes [11, 12].

Sex chromosomes

With the exception of the primitive family of ratites, male and female birds have clearly distinguishable sex chromosomes, indicating that the mechanism which controls the developmental decision to form an ovary or testis

is chromosomal. The avian karyotype is composed of both macrochromosomes and microchromosomes, the latter being very small in size and cytologically indistinguishable from one another. The chicken has 39 pairs of
chromosomes consisting of 10 pairs of macrochromosomes and 29 pairs of
microchromosomes [13]. The sex chromosomes of birds have been designated the Z chromosome and the W chromosome; males are homogametic
(ZZ) and females are heterogametic (ZW) [14, 15]. In the chicken, the
Z chromosome is the fifth largest of the chromosomes and comprises
approximately 7% of the genome, whereas the W chromosome is the size
of a microchromosome and comprises approximately 1.5% of the genome
[16, 17]. There is no significant homology between the avian Z and the
mammalian X chromosomes, suggesting that the sex chromosomes of
birds and mammals evolved from different pairs of autosomes [18, 19]. The
W chromosome does show some superficial similarity to the mammalian
Y chromosome, being composed largely of heterochromatic DNA, which
is made up mainly of two families of repeat sequences (Fig. 3) [20, 21]. In
addition, during meiosis, a small region at the tip of the short arm of the W
chromosome is able to pair with a similar region on the Z chromosome,
reminiscent of the mammalian pseudo-autosomal region (PAR) [22–24].
So far, only three genes, *CHD-W, DWM1/ATP5A1-W* and *ASW* have been
localized to the nonrepeat regions of the W chromosome. CHD-W is a
chromohelicase-DNA-binding protein [25], DMW1/ATP5A1-W is an avian homologue of the adenosine 5′-triphosphate (ADP) synthase α subunit
[19, 26], and both of these genes have related sequences on the Z chromosome. ASW is reported to be specific to the W chromosome and is of
unknown function [27]. The avian Z chromosome has been described as
composed of three discernible regions. There is a PAR at the tip of the short
arm, although no crossing over occurs in this region between the Z and W
chromosomes. The second region contains a single recombination nodule
located near the pairing end of the ZW bivalent, and the third region comprises the remainder of the short arm and the entire long arm, which is specific to the Z chromosome. Only a small number of genes and anonymous
DNA markers have so far been assigned to the chick Z chromosome; however, a concentrated mapping effort has recently been initiated, and a
chicken genome database established at the Roslin Institute [28]. Figure 3
illustrates the relative sizes of the chicken Z and W chromosomes and lists
the genes so far assigned to these.

 In mammals where the female has two X chromosomes and the male
only one, the imbalance of the copy number of X-linked genes is equalized
by regulation at the level of gene expression in the female [29]. This dosage
compensation is achieved by inactivating one of the female X chromosomes, which results in the cytological feature known as the Barr body
[30]. In birds, it has been widely accepted that there is no dosage compensation system in operation and that the level of expression of Z-linked
genes in males will be double that in females [13, 31]. Evidence cited in

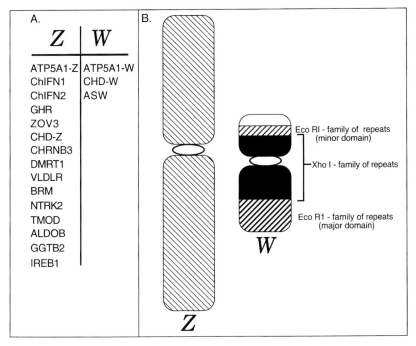

Figure 3. Sex chromosomes of the chicken. (*A*) Genes assigned to the Z and W chromosomes
of the chicken. ChIFN1: Interferon 1; ChIFN2: Interferon 2; ATP5A1: F1 ATP Synthase α;
ZOV3: Z-chromosome linked immunoglobulin superfamily; GHR: Growth Hormone Receptor;
CHD: Chromodomain Helicase DNA Binding Protein; CHRNB3: Nicotinic Acetylcholine
Receptor; DMRT1: DM domain containing transcription factor; VLDLR: very low density
lipoprotein receptor; BRM: Bromodomain containing protein; NTRK2: Neurotrophic Tyrosine
Kinase Receptor-Related; TMOD: Tropomodulin; ALDOB: Aldolase B; GGTB2: Glyco-
protein-4-beta Galactosyl Transferase; IREBP: Iron-Responsive Element Binding Protein;
ASW: Avian Sex-specific Gene W – chromosome [27, 28]. (*B*) Relative size of chicken sex
chromosomes and distribution of W chromosome repeat sequences.

support of this theory includes a failure to detect the presence of a Barr
body in males, no asynchronous replication of the Z chromosomes and a
report of higher levels of the Z-linked aconitase/IREB gene product in
males than females [32–34]. However, individually none of these data
would be conclusive proof that birds have no system of dosage compensation, and even in combination the evidence is mostly circumstantial. It is
still quite possible that a dosage-compensation system, not based on Z
chromosome inactivation, exists in birds.

Sex-determining mechanism

The mechanism which regulates the developmental decision to form a
testis or an ovary in birds is unknown, but there is wide support for two

different proposed mechanisms of sex determination. These are based on
models established in other organisms with heteromorphic sex chromosomes and propose that the sex-determining mechanism in birds is regulated by either a dominant genetic switch as employed by mammals or a
dosage-based mechanism as typified by the systems in operation in *Drosophila* and *Caenorhabditis elegans* [6, 15, 35, 36]. In mammals, the male-
specific gene Sry, located on the Y chromosome, has been identified as the
testis-determining gene [37, 38]. It has been demonstrated that expression
of this gene in the gonads at the appropriate time in development is necessary and sufficient to induce testis differentiation [39, 40]. In birds with
homogametic males (ZZ) and heterogametic females (ZW) a male-specific
testis-determining gene is inconceivable. Under the proposed "dominant
gene" model, the female-specific W chromosome would carry an ovary-
determining gene that would operate in a fashion analogous to the mammalian *Sry* gene. The alternative dosage mechanism would depend on the
ratio of Z chromosome number to the number of sets of autosomes. Under
this model, the male with two Z chromosomes and a diploid set of autosomes would have a ratio of 1.0, whereas the female with one Z chromosome would have a ratio of 0.5. Here the interaction of an autosomal factor
with a single dose (ZW) or a double dose (ZZ) of a Z-linked gene product
would decide the fate of the developing gonad.

In mammals, the nature of the sex-determining mechanism was elucidated as a direct result of identifying individuals with sex chromosome
abnormalities. Such individuals can arise due to the nondisjunction of
the sex chromosomes during oogenesis or spermatogenesis [41]. During
oogenesis, failure of the members of the X chromosome pair to separate
and move into one daughter cell will produce oocytes with either two X
chromosomes or none. Fertilization of such oocytes by normal spermatozoa will result in zygotes with either XXX, XO, XXY or YO sex chromosomes. Alternatively, nondisjunction during spermatogenesis can give rise
to sperm with no sex chromosomes, with both an X and a Y, or with two Y
sex chromosomes. Fertilization of a normal oocyte could then result in

Table 1. Theoretical sex chromosome genotypes and predicted
sex under the "dominant gene" and "Z : A ratio" models of sex
determination

Sex chromosomes	Predicted sex	
	dominant gene	Z : A ratio
ZO	male	female
ZZ	male	male
ZZZ	male	male
ZZW	female	male
ZW	female	female
WO	female	?

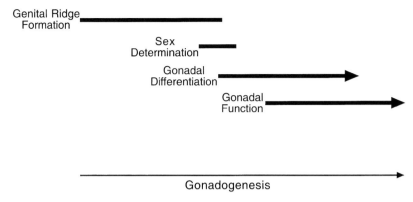

Figure 4. Phases of embryonic gonadal development.

zygotes with either XO, XXY or XYY sex chromosome complements. Klinefelter's syndrome (XXY or XXXY) and Turner's syndrome (XO) are relatively common examples of such occurrences [42]. The observation that in such cases individuals with a normal Y chromosome were always male and individuals lacking a Y chromosome were always female led to the conclusion that the Y chromosome carried a testis-determining gene. The identification of individuals with sex chromosome translocations and deletions finally led to the identification of the Sry gene as the testis-determining gene (reviewed in [43]).

The identification of birds with equivalent sex chromosome abnormalities would help to resolve which of the proposed models of sex determination was most likely correct. Table 1 list a number of potential avian genotypes and the predicted sex of each under both of the proposed models. Unfortunately, karyotype analysis has so far failed to reliably identify individuals with the truly informative genotypes such as ZO and ZZW. While a single report of a male ZZW aneuploid would appear to support the Z chromosome : autosome ratio model [44], others have convincingly argued that cytological techniques available at that time were not sufficiently reliable to assign a credible karyotype [45]. The most detailed karyotyping studies available in birds have resulted from the analysis of a triploid line of chickens by Thorne et al. [45–50]. In these studies, triploid birds with a ZZZ sex chromosome complement developed testes but were infertile, whereas ZZW birds were classified as intersexes with two developed gonads. In the ZZW triploids, the right gonad initially developed as a testis, whereas the left gonad resembled an ovary; however, both gradually became masculinized, and eventually both gonads produced abnormal spermatozoa. Table 2 summarizes the reliable aberrant karyotype information that is available for the chicken. While instances of individual karyotypes can be selected which support either of the proposed sex-determining models, overall there is insufficient evidence presently available to form a definite conclusion.

Table 2. Karyotypes report for the chicken and assigned phenotypes

Genotype	Phenotype	Predicted by model	
		dominant gene	Z : A ratio
2A : ZW	female	yes	yes
2A : ZZ	male	yes	yes
3A : ZZZ	male	yes	yes
3A : ZZW	intersex	no	yes
3AZWW	female	yes	yes

Agreement of individual karyotypes with the "dominant gene" and "Z : A ratio" models of sex determination is indicated. (3AZWW karyotype was reported for a single individual sexed as female at day 16 of embryonic development.)

Genes in gonadal development

The embryonic development of the gonad can be considered as a series of sequential but overlapping phases, as depicted in Figure 4. In mammals, a number of the genes associated with the development of these phases have been identified, and these are listed in table 3 [37, 38, 51–57].

We and others have attempted to isolate chick homologues to the genes involved in mammalian gonadal development and to study the expression of these genes during gonadal development in the chick. Attempts to identify a direct homologue to the mammalian testis-determining gene (Sry) have so far proved unsuccessful, although a large number of related SOX (Sry box) genes have been identified [58–60]. Given that there is no male-specific sex chromosome in birds, the failure to detect a Sry homologue may be unsurprising. However, the possibility that a closely related SOX gene on the W chromosome may act as an ovary-determining gene cannot be excluded, and we have demonstrated that a number of SOX genes are expressed in the genital ridge at the appropriate time in development [58]. With the exception of Sry, and to date Dax-1, chick homologues to the remaining genes listed in Table 3 have been isolated [57, 61–64]. The expression of the majority of these genes during gonadal development has

Table 3. Association of previously identified genes with specific phases of gonadal development

Genital ridge formation	Sex determination	Gonadal differentiation	Gonadal function
WT 1	Sry	SF 1	AMH
SF 1		Sox9	steroid-metabolizing
		Dax 1	enzymes (including
			aromatase)

WT-1, Wilms's tumor gene-1; SF-1, steroidogenic factor-1; Sox9, Sry box-containing gene 9; Sry, sex-determining region Y chromosome; Dax-1, DSS-AHC critical region on the X chromosome, gene 1; AMH, anti-Müllerian hormone [38–40, 51–55, 57].

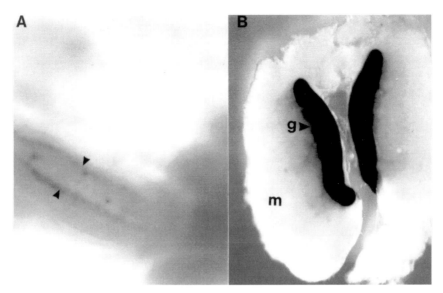

Figure 5. Expression of SF-1 and AMH in developing chick gonads by whole-mount *in situ* hybridization analysis. (*A*) Expression of SF-1 in the chick embryo at 3.5 days of development. The embryo has been decapitated and the ventral surface and viscera removed. Hybridization was performed using a digoxygenin-labeled antisense SF-1 probe. The signal is indicated by purple deposits. The genital ridge is indicated by arrowheads. (*B*) Expression of AMH in the isolated gonads and mesonephroi of the chick embryo at day 8 of incubation. Hybridization was performed using a digoxygenin-labeled antisense AMH probe. g, gonad; m, mesonephros.

been analyzed by various groups [57, 63, 65–69]. While these results indicate similar patterns of expression in birds and mammals, the piecemeal nature of these reports makes it difficult to ascertain the exact temporal relationships of the expression patterns. We have prepared RNA populations from pools of male gonads and pools of female gonads isolated from embryos incubated for 4.5, 5.5, 6.5, 7.5 and 8.5 days. We have analzyed the expression of the chick, WT-1, SF-1, Sox9, AMH and aromatase genes in these RNA samples by Northern analysis and in isolated gonads by *in situ* hybridization (unpublished data). Our data reveal that WT-1, Sox9 and SF-1 are expressed in the developing gonad of both sexes throughout this period, with Sox9 and SF-1 being expressed at higher levels in males than females from day 6.5 onwards. AMH expression is first detectable at very low levels in male gonads at day 5.5 of development and at increasing levels thereafter (Fig. 5 shows examples of SF-1 and AMH expression).

While the embryonic expression of AMH in mammals is male-specific, in birds AMH is required to effect the regression of the right Müllerian duct, and consequently a low level of AMH expression is detectable in both right and left female gonads from day 6.5 onwards. Aromatase is required for the conversion of androgens to estrogen, and the expression of the chick

Table 4. Comparison of expression profiles of genes associated with gonadal development in birds and mammals

Genes	Mammals	Birds
WT-1	Expressed in males and females throughout gonadal development.	Expressed in males and females throughout gonadal development.
SF-1	Expressed in males and females throughout gonadal development. Expressed at higher levels in males during differentiation phase.	Expressed in males and females throughout gonadal development. Expressed at higher levels in males during differentiation phase.
Sox9	Expressed in males and females during sex determination. Male-specific expression during differentiation phase.	Expressed in males and females during sex determination. Expressed at higher levels in males during differentiation phase.
Sry	Male-specific expression during sex-determination phase.	Absent.
Dax-1	Expressed in males and females during differentiation phase. Higher levels of expression in females (*).	Unknown.
AMH	Male-specific expression during differentiation and function phases.	Expressed in both males and females during differentiation and function phases. Higher levels in males.
Aromatase	Expressed in males and females during late differentiation and function phases.	Female-specific expression during late differentiation and function phases.

Expression profiles in the chicken may not reflect gonad-specific expression, as profiles are derived from Northern analyses of RNA isolated from genital ridge plus mesonephros (unpublished data). *In situ* hybridization studies report male-specific expression of Sox9 during the differentiation phase in the chicken [56, 69]. *Recent studies reports higher levels of Dax-1 in males during the differentiation phase [95] and suggest that Dax-1 has a critical role in spermatogenesis rather than ovary determination [96]. *WT-1*, Wilms's tumor gene-1; *SF-1*, steroidogenic factor-1; *Sox9*, Sry box-containing gene 9; Sry, sex-determining region Y chromosome; *Dax-1*, DSS-AHC critical region on the X chromosome, gene 1; AMH, anti-Müllerian hormone.

aromatase gene is detectable from day 6.5 only in female gonads. The expression patterns that we observed in this study are in general agreement with patterns reported for these individual genes by others [57, 63, 65–69]. The expression profiles of WT-1, SF-1 and Sox9 suggest a high degree of conservation of the molecular mechanisms regulating gonadal development in the chick and mammal. The temporal relationships between the expression profiles of these genes and those of *AMH* and *aromatase* would indicate that sex determination in chickens occurs no later than day 5.5 of embryonic development. A comparison of the expression profiles of the major genes involved in gonadal development in birds and mammals is shown in Table 4.

Candidate sex determining genes

Theoretically at least, all the genes so-far assigned to the avian sex chromo-somes are candidate sex determining genes under either the "dominant gene" or "Z:A ratio" models of sex determination. In reality, the majority of avian sex chromosome genes can be discounted as serious candidate sex determining genes. This is because there is either insufficient evidence (e.g. expression into functional protein has not been demonstrated) or the gene (or homologue in other species) already has a well-documented function unrelated to sex determination. However, one gene which has recently been mapped to the chicken Z-chromosome (*DMRT1*) [97], does qualify as a serious candidate sex determining gene. In humans it is thought that DMRT1 operates in a dose-dependent fashion as a testis determinant and it is known that deletions of the region of chromosome 9 containing *DMRT1* are associated with some cases of gonadal dysgenesis and XY sex reversal [98]. It is also interesting to note that human *DMRT1* was orig-inally identified due to sequence similarity with genes known to regulate the sexual development of invertebrates, *doublesex* and *mab-3*. While DMRT1 expression in birds remains to be documented, without dosage compensation, males would be expected to have double the dose seen in females. If this were the case and if expression was confined to the gonads, it would represent strong circumstantial evidence for a major role for DMRT1 in the sexual development of birds.

The role of steroids in avian gonadal development

Shortly after the onset of gonadal differentiation, the embryonic gonads of both birds and mammals are considered to have the capacity to produce steroids [70, 71]. In mammals, steroids are thought to be responsible for the development of accessory sex organs and secondary sex characteristics and to play no role in the development of the gonads [72]. However, in birds it appears that steroids produced by the embryonic gonad may play a major role in the development of the gonads themselves. This evidence comes from a large body of work focused on sex reversal which has been carried out in the chicken. In addition to documenting cases of spontaneous sex reversal, experimental approaches have included castration, transplanta-tion, administration of various compounds including estrogens and andro-gens, antiestrogens and antiandrogens, and aromatase inhibitors (a non-exhaustive list includes [73–79]). From these data it would appear that both male and female sex reversal can be induced under certain experi-mental conditions, although in cases of spontaneous sex reversal these are most commonly female to male. Experiments have shown that if the synthesis of estrogen is blocked in early genetically female embryos by the administration of an aromatase inhibitor, approximatley 50% of the embryos develop testis [80]. Similarly, embryonic testicular implants can cause testis development in genetically female embryos, and this reversal is preventable by the simultaneous administration of estrogens [81]. Conversely, the administration of exogenous estrogen can feminize the gonads of male embryos, whereas estrogen antagonists are able to disrupt the normal development of the ovary [49, 82, 83]. From these reports estrogens would appear to be an absolute requirement for the development of the normal ovary. This ability of steroids to sex-reverse the embryonic gonads implies a certain "plasticity" in the development of the avian gonads which is not apparent in the development of the mammalian gonad. Reports of sex reversal of adult animals would also suggest that this plasticity is not confined to the embryonic stages. However, the majority of these cases seem to be the result of either ovarian disease, which presumably alters the steroid profile produced, or the effects of environmental steroids. It seems likely that reports of adult sex reversal are the result of hormonal influence on secondary sexual characteristics rather than gonadal transformation.

The ability of steroids to effect physiological changes is often the result of modifications in the expression patterns of specific genes and is mediated through specific nuclear proteins, the steroid receptors. These receptors belong to a large family of ligand-activated transcription factors which regulate gene expression by interacting either in a protein/DNA manner with cognate DNA sequences called responsive elements [84] or in a protein/protein manner with other transcription factors [85]. Recent reports of the expression of the estrogen receptor gene in chicken have further emphasized the importance of estrogen in chick gonadal development [68, 86, 87]. These reports demonstrated that the expression of the estrogen receptor gene is gonad-specific in the early embryo and detail the expression of this gene throughout gonad development. Using *in situ* hybridization and reverse transcriptase-polymerase chain reaction (RT-PCR), estrogen receptor expression was detected in the gonads, coincident with the first indications of differentiation. The authors reported estrogen receptor expression in both gonads of male and female embryos with higher levels of expression in the left gonads. The expression of the estrogen receptor in the male gonads was transitory but could be maintained by the administration of estrogen. The expression data on the estrogen receptor coupled with the female-specific ability to synthesize the appropriate ligand strongly supports a pivotal role for estrogen in regulating the development of the ovary. Expression of the estrogen receptor in the developing male gonads would also explain the ability of exogenous estrogen to induce ovary formation in male embryos. Slightly at variance with this suggestion are the results of immunohistochemistry studies by Andrews et al. [86] which detected the presence of the estrogen receptor protein only in female gonads.

While the weight of evidence would point to a crucial role for estrogen in inducing cortex formation in the differentiating gonad, data from Andrews et al. [86] and Smith et al. [68] also raise the possibility that

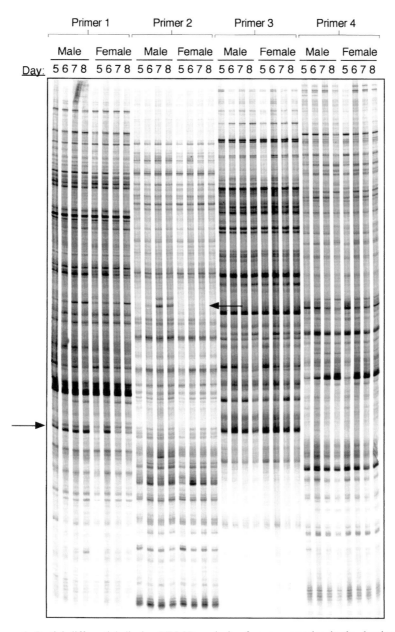

Figure 6. Partial differential display RT-PCR analysis of gene expression in the developing gonads of male and female chick embryos. Gel bands represent individual gene transcripts, and each primer combination displays ~ 250 transcripts at each stage of development. Examination of display gel allows a comparison of the expression profiles of genes in male and females throughout gonadal development. Genes differentially expressed in the male and female gonads are readily identified as indicated by arrows. Day, day of development.

estrogen and/or the estrogen receptor may play an even more central role in the sex-determining process. Although Nakabayashi et al. [87] were unable to detect expression of the estrogen receptor gene until after the point of sex determination, Andrews et al. [86] and Smith et al. [68] report expression of this transcript in both male and female gonads continuously from day 4.5 to day 12.5 of incubation. In addition Smith et al. [68] report a sexually dimorphic pattern of expression of the estrogen receptor in embryos at day 3.5 of incubation. In these embryos, estrogen receptor gene expression was confined to the genital ridge of female embryos only. This raises the possibility that estrogen and the estrogen recetor could play an important role in the formation of the genital ridge and in very early sex-determining events. Studies reporting the surprising detection of steroids in the genital ridge of very early chick embryos would appear to reinforce this possibility [88, 89]. These studies reported the presence of steroids in day 3.5 embryos with higher levels of estrogens in females and higher levels of androgens in males by day 5.5, providing a mechanism whereby steroids could regulate sex specific differentiation of the testis and ovary from the time of genital ridge formation. Unfortunately, it is hard to reconcile these reports with evidence on the expression patterns of genes coding for the steroid-metabolizing enzymes [65–68]. The expression of these genes, indicative of the differentiation of the specialized cells of the testis and ovary, cannot be detected until day 6.5 of incubation. In an attempt to clarify this issue, we have used radioimmunoassay to measure estradiol 17β levels in the gonads of male and female chick embryos from day 4.5 to day 12.5 of development (unpublished data). In contrast to the earlier studies, we were unable to detect estrogens in the male gonads at any stage and only from day 6.5 onwards in the female gonad. Our observations on estrogen synthesis are in complete agreement with the coincident expression of the aromatase gene reported by ourselves and others [65–68]. Despite the apparent conflict between the presence of the receptor and the availability of the ligand, the possibility remains that the estrogen receptor may regulate gene expression via an as yet unidentified alternative ligand or by ligand-free transactivation, and could still play a central role in formation of the genital ridge and sex determination in birds.

Future prospects

Recent studies suggest that many of the genes involved in the differentiation of the specialized cells of the mammalian gonad are likely to play similar roles during avian gonadogenesis. In addition, these studies have also helped to understand some of the tremendous legacy of data on sex reversal in birds. Despite these advances, the nature of the chromosomal mechanism by which sex is determined in birds remains unknown. While shadowing advances in the mammalian field has so far proved fruitful, it is unlikely that the approaches used to elucidate the sex determining mechanism and identify the testis-determining gene in mammals will succeed in birds. This is principally due to the difficulties in identifying individuals with chromosomal abnormalities and the lack of detailed information on the sex chromosomes. In addition, it appears as if the process by which the sex-determining mechanism is implemented in birds may be more akin to that in fish and reptiles where estrogen plays a central role [90, 91] than to that in mammals. For these reasons, a number of groups have invested considerable effort in alternative approaches such as differential display – RT-PCR [92, 93] and representational difference analysis [94]. These approaches compare gene expression in the developing male and female gonads, with the aim of identifying transcripts which exhibit sex-specific or sexually dimorphic expression. A typical example of a differential display analysis is shown in Figure 6. A notable success using this approach was the identification of the female-specific *ASW* gene [27], and we have recently isolated a Z-linked gene which is expressed in the developing gonads with higher levels in males than in females (unpublished data).

The rapidly expanding Chickmap program and recent improvements in techniques such as differential display should help to identify the genes involved in avian sex determination and gonadal development. However, to fully understand these processes and to assess the significance of novel genes, a concentrated effort is required to identify individuals with informative sex chromosome genotypes and to resolve the issue of dosage compensation.

Acknowledgement
This work was supported by a commission from the Ministry of Agriculture, Fisheries and Food, United Kingdom.

References

1 Browder LW, Erickson CA and Jeffery WR (1991) Organogenesis: gonad development and sex differentiation. In: Developmental Biology, pp. 661–683, Browder LW, Erickson CA and Jeffery WR (eds), Saunders College Publishing, Philadelphia
2 Romanoff AL (1960) The Avian Embryo, Macmillan, New York
3 Martineau J, Nordqvist K, Tilmann C, Lovell-Badge R and Capel B (1997) Male-specific cell migration into the developing gonad. *Curr Biol* 7: 958–968
4 Pelliniemi LJ, Frojdman K, Sundstrom J, Pollanen P and Kuopio T (1998) Cellular and molecular changes during sex differentiation of embryonic mamalian gonads. *J Exp Zool* 281: 482–494
5 Gilbert AB (1979) Female Genital Organs, Academic Press, London
6 McCarrey JR and Abbott UK (1979) Mechanisms of genetic sex determination, gonadal sex differentiation and germ-cell development in animals. *Adv Genet* 20: 217–289
7 Maraud R, Vergnaud O and Reshedi M (1987) Structure of the right testis of sexually mature genetically female fowl experimentally masculinized during embryonic life and submitted to a posthatching left castration. *Gen Comp Endocrinol* 68: 208–215
8 Rashedi M, Maraud R and Stoll R (1983) Development of the testes in female domestic-fowls submitted to an experimental sex reversal during embryonic life. *Biol Reprod* 29: 1221–1227

 9 Groenendijk-Huijbers MM (1965) The right ovary of the chick embryo after early sinistral castration. *Anat Rec* 153: 93–106
10 Groenendijk-Huijbers MM (1973) Influence of gonadal hormone administration and gonadal implantation on the compensatory growth of the embryonic chicken right ovary after early sinsitral castration. *Verh Anat Gest* 67: 193–196
11 Witschi E (1935) Origin of asymmetry in the reproductive system of birds. *Am J Anat* 56: 119–141
12 Van Krey HP (1990) Reproductive biology in relation to breeding and genetics. In: Poultry Breeding and Genetics, Crawford RD (ed), Elsevier, Amsterdam
13 Schmid M, Enderle E, Schindler D and Schempp W (1989) Chromosome-banding and DNA-replication patterns in bird karyotypes. *Cytogenet Cell Genet* 52: 139–146
14 Bloom SE (1974) Current knowledge about the avian W chromosome. *Bioscience* 24: 340–344
15 Bitgood J and Shoffner RN (1990) Cytology and cytogenetics. In: Poultry Breeding and Genetics, pp. 401–427, Crawford RD (ed), Elsevier, Amsterdam
16 Fechheimer NS (1990) Chromosomes of Chickens, Academic Press, London
17 Tone M, Sakaki Y, Hashiguchi T and Mizuno S (1984) Genus specificity and extensive methylation of the W-chromosome-specific repetitive DNA-sequences from the domestic-fowl, *Gallus-gallus-domesticus. Chromosoma* 89: 228–237
18 Graves JAM and Reed KC (1993) Organization, function and evolution of the sex chromosomes. In: Sex Chromosomes and Sex Determining Genes, pp. 129–135, Graves JAM and Reed KC (eds), Harwood Academic Publishers, Australia
19 Fridolfsson AK, Cheng H, Copeland NG, Jenkins NA, Liu HC, Raudsepp T et al (1998) Evolution of the avian sex chromosomes from an ancestral pair of autosomes. *PNAS* 95: 8147–8152
20 Tone M, Nakano N, Takao E, Narisawa S and Mizuno S (1982) Demonstration of W-chromosome-specific repetitive DNA-sequences in the domestic-fowl, *Gallus g. domesticus. Chromosoma* 86: 551–569
21 Saitoh Y, Saitoh H, Ohtomo K and Mizuno S (1991) Occupancy of the majority of DNA in the chicken W-chromosome by bent-repetitive sequences. *Chromosoma* 101: 32–40
22 Solari AJ, Fechheimer NS and Bitgood JJ (1988) Pairing of ZW gonosomes and the localized recombination nodule in 2 Z-autosome translocations in *Gallus domesticus. Cytogenet Cell Genet* 48: 130–136
23 Solari AJ, Thorne MH, Sheldon BL and Gillies CB (1991) Synaptonemal complexes of triploid (ZZW) chickens – Z-Z pairing predominates over Z-W pairing. *Genome* 34: 718–726
24 Solovei I, Gaginskaya E, Allen T and Macgregor H (1992) A novel structure associated with a lampbrush chromosome in the chicken, *Gallus-domesticus. J Cell Sci* 101: 759–772
25 Griffiths R and Korn RM (1997) A CHD1 gene is Z chromosome linked in the chicken *Gallus domesticus. Gene* 197: 225–229
26 Dvorák J, Halverson JL, Gulick P, Rauen KA, Abbott UK, Kelly BJ et al (1992) cDNA cloning of a Z- and W-linked gene in gallinaceous birds. *J Hered* 83: 22–25
27 O'Neill MJ and Sinclair AH (1997) Identification of novel sex specific transcripts in the chick genital ridge by Representation Difference Analysis. Ist International Symposium on the Biology of Vertebrate Sex Determination, May 1997, p. 43, Hawaii
28 ChickMap. A physical and genetic map of the chick genome (1998) hhtp://www.ri.bbsrc.ac.uk/chickmap/
29 Migeon RB (1994) X-chromosome inactivation – molecular mechanisms and genetic consequences. *Trends Genet* 10: 230–235
30 Lyon MF (1989) X-chromosome inactivation as a system of gene dosage compensation to regulate gene-expression. *Prog Nucleic Acid Res Mol Biol* 36: 119–130
31 Chandra SH (1994) Proposed role of W chromosome inactivation and the absence of dosage compensation in avian sex determination. *Proc R. Soc Long B Biol Sci* 258: 79–82
32 Bianchi NO and Molina OJ (1967) Chronology and pattern and replication in the bone marrow chromosomes of *Gallus domesticus. Chromosoma* 21: 387–397
33 Takagi N (1972) A comparitive study of the chromosome replication in six species of birds. *Jpn J Genet* 47: 115–123
34 Baverstock PR, Adams M, Polkinghorne RW and Gelder M (1982) A sex-linked enzyme in birds-Z – chromosome conservation but no dosage compensation. *Nature* 296: 763–766

35 Sittmann K (1984) Sex determination in birds – progeny of nondisjunction canaries of Durham. *Genet Res* 43: 173–180

36 Halverson JL and Dvorak J (1993) Genetic-control of sex determination in birds and the potential for its manipulation. *Poult Sci* 72: 890–896

37 Gubbay J, Collignon J, Koopman P, Capel B, Economou A, Munsterberg A et al (1990) A gene-mapping to the sex-determining region of the mouse Y-chromosome is a member of a novel family of embryonically expressed genes. *Nature* 346: 245–250

38 Sinclair AH, Berta P, Palmer MS, Hawkins JR, Griffiths BL, Smith MJ et al (1990) A gene from the human sex-determining region encodes a protein with homology to a conserved DNA-binding motif. *Nature* 346: 240–244

39 Koopman P, Munsterberg A, Capel B, Vivian N and Lovell-Badge R (1990) Expression of a candidate sex-determining gene during mouse testis differentiation. *Nature* 348: 450–452

40 Koopman P, Gubbay J, Vivian N, Goodfellow P and Lovell-Badge R (1991) Male development of chromosomally female mice transgenic for SRY. *Nature* 351: 117–121

41 Langman J (1975) Urogenital system. In: Medical Embryology, pp. 160–200, Langman J (ed), Williams and Wilkins, Baltimore

42 Short RV (1972) Sex determination and differentiation. In: Reproduction in Mammals, Vol. 2, pp. 43–71, Austin CR and Short RV (eds), Cambridge University Press

43 Wachtel SS and Tiersch TR (1994) The search for the male-determining gene. In: Molecular Genetics of Sex Determination, pp. 1–22 Wachtel SS (ed), Academic Press, San Diego

44 Crew FAE (1933) A cas eof non-disjunctionin the fowl. *Proc R Soc Edin* 53: 89–105

45 Thorne MH (1995) Genetics of poultry reproduction. In: World Animal Science Poultry Production, pp. 411–434, Hounton P (ed), Elsevier, Amsterdam

46 Thorne MH and Sheldon BL (1993) Triploid intersex and chimeric chickens: Useful models for studies of avian sex determination. In: Sex Chromosomes and Sex-Determining Genes, pp. 199–205, Reed KC and Graves JAM (eds), Harwood Academic Publishers, Australia

47 Thorne MH and Sheldon BL (1991) Cytological evidence of maternal meiotic errors in a line of chickens with a high-incidence of triploid. *Cytogenet Cell Genet* 57: 206–210

48 Thorne MH, Collins RK and Sheldon BL (1991) Triploidy and other chromosomal-abnormalities in a selected line of chickens.*Genet Select Evol* 23: S212–S216

49 Thorne MH, Martin ICA, Lin M and Sheldon BL (1992) Hormone induced sex change in triploid fowl. Proceedings of the Fourteenth World Poultry Congress, September 1997, Amsterdam, 788–791

50 Lin M, Thorne MH, Martin ICA, Sheldon BL and Jones RC (1995) Development of the gonads in the triploid (ZZW and ZZZ) fowl, *Gallus-domesticus*, and comparison with normal diploid males (ZZ) and females (ZW). *Reprod Fertil Dev* 7: 1185–1197

51 Luo XR, Ikeda Y and Parker KL (1994) A cell-specific nuclear receptor is essential for adrenal and gonadal development. *Cell* 77: 481–490

52 Foster JW, Dominguezsteglich MA, Guioli S, Kwok C, Weller PA, Stevanovic M et al (1994) Campomelic dysplasia and autosomal sex reversal caused by mutations in an SRY-related gene. *Nature* 372: 525–530

53 Swain A, Zanaria E, Hacker A, Lovel-Badge R and Camerino G (1996) Mouse Dax1 expression is consistent with a role in sex determination as well as in adrenal and hypothalamus function. *Nature Genet* 12: 404–409

54 Josso N and Picard JY (1986) Anti-mullerian hormone. *Physiol Rev* 66: 1038–1090

55 Wagner T, Wirth J, Meyer J, Zabel B, Held M, Zimmer J et al (1994) Autosomal sex reversal and campomelic dysplasia are caused by mutations in and around the SRY-related gene SOX9. *Cell* 79: 1111–1120

56 Kent J, Wheathley SC, Andrews JE, Sinclair AH and Koopman P (1996) A male-specific role for SOX9 in vertebrate sex determination. *Development* 122: 2813–2822

57 Kent J, Coriat AM, Sharpe PT, Hastie ND and Vanheyningen V (1995) The evolution of WT1 sequence and expression pattern in the vertebrates. *Oncogene* 11: 1781–1792

58 McBride D, Sang H and Clinton M (1997) Expression of Sry-related genes in the developing genital ridge/mesonephros of the chick embryo. *J Reprod Fertil* 109: 59–63

59 Coriat A-M, Muller U, Harry JL, Uwanogho D and Sharpe PT (1993) PCR amplification of SRY-related gene sequences reveals evolutionary conservation of the SRY-box motif. *PCR Meth Appl* 2: 218–222

60 Griffiths R (1991) The isolation of conserved DNA-sequences related to the human sex-determining region Y-gene from the lesser black-backed gull (*Larus-fuscus*). *Proc R Soc Lond B Biol Sci* 244: 123–128

61 Kudo T and Sutou S (1997) Molecular cloning of chicken FTZ-F1-related orphan receptors. *Gene* 197: 261–268

62 Uwanogho D, Rex D, Cartwright EJ, Pearl G, Scotting PJ and Sharpe PT (1994) Direct submission-GenBank. *Accession number* U12533

63 Eusebe DC, Diclemente N, Rey R, Pieau C, Vigier B, Josso N et al (1996) Cloning and expression of the chick anti-mullerian hormone gene. *J Biol Chem* 271: 4798–4804

64 McPhaul MJ, Noble JF, Simpson ER, Mendelson CR and Wilson JD (1988) The expression of a functional cDNA encoding the chicken cytochrome p-450 arom (aromatase) that catalyzes the formation of estrogen from androgen. *J Biol Chem* 263: 16358–16363

65 Yoshida K, Shimada K and Saito N (1996) Expression of p450(17-alpha) hydroxylase and p450 aromatase genes in the chicken gonad before and after sexual-differentiation. *Gen Comp Endocrinol* 102: 233–240

66 Clinton M (1998) Sex determination and gonadal development: a bird's eye view. *J Exp Zool* 281: 457–465

67 Shimada K (1998) Gene expression of steroidogenic enzymes in chicken embryonic gonads. *J Exp Zool* 281: 450–456

68 Smith CA, Andrews JE and Sinclair AH (1997) Gonadal sex differentiation in chicken embryos: expression of estrogen receptor and aromatase genes. *J Steroid Biochem Mol Biol* 60: 295–302

69 daSilva SM, Hacker A, Harley V, Goodfellow P, Swain A and Lovell-Badge R (1996) Sox9 expression during gonadal development implies a conserved role for the gene in testis differentiation in mammals and birds. *Nature Genet* 14: 62–68

70 Greco TL and Payne AH (1994) Ontogeny of expression of the genes for steroidogenic enzymes p-450 side-chain cleavage, 3b-hydroxysteroid dehydrogenase, p-450 17a-hydroxylase/C_{17-20} lyase, and p-450 aromatase in fetal mouse gonads. *Endocrinology* 135: 262–268

71 Sweeney T, Saunders PTK, Millar MR and Brooks AN (1997) Ontogeny of anti-mullerian hormone, 3 beta-hydroxysteroid dehydrogenase and androgen receptor expression during ovine fetal gonadal development. *J Endocrinol* 153: 27–32

72 Crews D (1994) Temperature, steroids and sex determination. *J Endocrinol* 142: 1–8

73 Abdel-Hameed F and Shoffner RN (1971) Intersexes and sex determination in chicken. *Science* 172: 962–964

74 Abinawanto K, Shimada K, Yoshida K and Saito N (1996) Effects of aromatase inhibitor on sex-differentiation and levels of p450(17-alpha) and p450(arom) messenger-ribonucleic-acid of gonads in chicken embryos. *Gen Comp Endocrinol* 102: 241–246

75 Frankenhuis MT and Kappert HJ (1980) Experimental transformation of right gonads of female fowl into fertile testis. *Biol Reprod* 23: 526–529

76 Hutson JM, Donahoe PK and Maclaughlin DT (1985) Steroid modulation of mullerian duct regression in the chick-embryo. *Gen Comp Endocrinol* 57: 88–102

77 Maraud R and Vergnaud O (1986) Development of interstitial-cells in experimentally sex-reversed gonads of genetically female chick-embryos. *Gen Comp Endocrinol* 63: 464–470

78 Maraud R, Rashedi M and Stoll R (1986) Experimentally induced true hermaphroditism in genetically female fowl. *Rouxs Arch Dev Biol* 195: 10–14

79 Wartenberg H, Lenz E and Schweikert H-U (1992) Sexual differentiation and the germ cell in sex reversed gonads after aromatase inhibition in the chicken embryo. *Andrologia* 24: 1–6

80 Elbrecht A and Smith RG (1992) Aromatase enzyme-activity and sex determination in chickens. *Science* 255: 467–470

81 Stoll R, Rashedi M and Maraud R (1982) Action of estradiol on the hermaphroditism induced by testicular graft in the female chick-embryo. *Gen Comp Endocrinol* 47: 190–199

82 Stoll R, Ichas F, Faucounau N and Maraud R (1993) Action of estradiol and tamoxifen on the testis-inducing activity of the chick embryonic testis grafted to the female embryo. *Anat Embryol* 188: 587–592

83 Perrin FMR, Stacey S, Burgess AMC and Mittwoch U (1995) A quantitative investigation of gonadal feminization by diethylstilbestrol of genetically male embryos of the quail *Coturnix-coturnix-japonica*. *J Reprod Fertil* 103: 223–226

84 Parker MG (1991) Nuclear Hormone Receptors: Molecular Mechanisms, Cellular Functions, Clinical Abnormalities, Academic Press, London
85 Gaub MP, Bellard M, Scheuer I, Chambon P and Sassonecorsi P (1990) Activation of the ovalbumin gene by the estrogen-receptor involves the fos-jun complex. *Cell* 63: 1267–1276
86 Andrews JE, Smith CA and Sinclair AH (1997) Sites of estrogen receptor and aromatase expression in the chicken embryo. *Gen Comp Endocrinol* 108: 182–190
87 Nakabayashi O, Kikuchi H, Kikuchi T and Mizuno S (1998) Differential expression of genes for aromatase and estrogen receptor during the gonadal development in chicken embryos. *J Mol Endocrinol* 20: 193–202
88 Woods JE and Podczaski ES (1974) Androgen synthesis in the gonads of the chick embryo. *Gen Comp Endocrinol* 24: 413–423
89 Woods JE and Erton LH (1978) The synthesis of estrogen in the gonads of the chick embryo. *Gen Comp Endocrinol* 36: 360–370
90 Lance VA and Bogart MH (1992) Disruption of ovarian development in alligator embryos treated with an aromatase inhibitor. *Gen Comp Endocrinol* 86: 59–71
91 Piferrer F, Zanuy S, Carrillo M, Solar II, Devlin RH and Donaldson EM (1994) Brief treatment with an aromatase inhibitor during sex differentiation causes chromosomally female salmon to develop as normal, functional males. *J Exp Zool* 270: 255–262
92 Liang PP (1992) Differential display of eukaryotic messanger RNA by means of the polymerase chain reaction. *Science* 257: 967–971
93 Miele G, MacRae L, McBride D, Manson J and Clinton M (1998) Elimination of false positives generated through PCR re-amplification of differential display cDNA. *Biotechniques* 25: 138–144
94 Lisitsyn NA, Lisitsyn N and Wigler M (1993) Cloning the differences between two complex genomes. *Science* 259: 946–951
95 Nachtigal WM, Hirokawa Y, Enyeart-VanHouten DL, Flanagan JN, Hammer GD and Ingraham HA (1998) Wilms' tumor 1 and dax-1 modulate the orphan nuclear receptor SF-1 in sex specific gene expression. *Cell* 93: 445–454
96 Yu RN, Ito M, Saunders TL, Camper SA and Jameson JL (1998) Role of Ahch in gonadal development and gametogenesis. *Nature Genet* 20: 353–357
97 Nanda I, Shan Z, Schartl M, Burt DW, Koehler M, Nothwang H-G et al (1999) 300 million years of conserved synteny between chicken Z and human chromosome 9. *Nature Genet* 21: 258–259
98 Raymond CS, Parker ED, Kettlewell JR, Brown LG, Page DC, Kusz K et al (1999) A region of human chromosome 9 required for testis development contains two genes related to known sexual regulators. *Hum Mol Genet* 8: 989–996

Genes and Mechanisms in Vertebrate Sex Determination
ed. by G. Scherer and M. Schmid
© 2001 Birkhäuser Verlag Basel/Switzerland

Temperature-dependent sex determination and gonadal differentiation in reptiles

Claude Pieau, Mireille Dorizzi and Noëlle Richard-Mercier

Institut Jacques Monod, CNRS, and Universités Paris 6 et Paris 7, 2 Place Jussieu, F-75251 Paris Cedex 05, France

Summary. In many reptile species, sexual differentiation of gonads is sensitive to temperature (temperature-dependent sex determination, TSD) during a critical period of embryonic development (thermosensitive period, TSP). Experiments carried out with different models including turtles, crocodilians and lizards have demonstrated the implication of estrogens and the key role played by aromatase (the enzyme complex that converts androgens to estrogens) in ovary differentiation during TSP and in maintenance of the ovarian structure after TSP. In some of these experiments, the occurrence of various degrees of gonadal intersexuality is related to weak differences in aromatase activity, suggesting subtle regulations of the *aromatase* gene at the transcription level. Temperature could intervene in these regulations. Studies presently under way deal with cloning (cDNAs) and expression (mRNAs) of genes that have been shown, or are expected, to be involved in gonadal formation and/or differentiation in mammals. Preliminary results show that homologues of the *WT1, SF1, SOX9, DAX1* and *AMH* genes exist in TSD reptiles. However, the expression patterns of these genes during gonadal differentiation may be different between mammals and TSD reptiles and also between different reptile species. How these genes could interact with aromatase is being examined.

Introduction

As fish and amphibians, reptiles exhibit different mechanisms of sex determination. In some species including snakes, many lizards and a minority of turtles, sex is determined by a gene or genes carried on sex chromosomes, according to male (XY/XX) or female (ZW/ZZ) heterogamety. The ZW/ZZ mechanism only exists in snakes, whereas both XY/XX and ZW/ZZ mechanisms are found in lizards and turtles. In this genotypic (GSD, [1, 2]), also called chromosomal sex determination (CSD, [3]), sex chromosomes vary from strongly heteromorphic (as in Viperidae) to slightly or not heteromorphic (as in Boidae and turtles). In other reptiles, all oviparous, sexual differentiation of gonads is sensitive to the incubation temperature of the eggs during a critical period of embryonic development.

In 1966, biased sex ratios related to different conditions of egg incubation (substrate, temperature) were reported in a lizard, *Agama agama*, suggesting an influence of temperature [4].

In 1971–72, gonadal differentiation was shown to be temperature-sensitive in the European freshwater turtle *Emys orbicularis* and the Mediterranean tortoise *Testudo graeca* [5, 6]. Later, the study of the effects of a wide range of temperatures demonstrated the influence of egg incubation temperature on the hatchling sex ratio also in an American turtle, *Chelydra serpentina* [7]. As of the end of the 1970s, these pioneer works were extended to many species and showed that the so-called temperature-dependent sex determination (TSD, [1, 2]) is widespread in reptiles. Thus TSD has been found in all crocodilians studied so far (i.e. half of the extant species), most turtles, some lizards [8] and more recently in the two living closely related species of *Sphenodon* [9].

In TSD reptiles, masculinizing temperatures yield 100% or a majority of males, whereas feminizing temperatures yield 100% or a majority of females. In the transition range(s) of temperature (TRT), males and females and sometimes intersexes are obtained. However, the responses to incubations at different constant temperatures vary according to three different patterns. In many turtles including all sea turtles, temperatures below TRT are masculinizing, whereas those above TRT are feminizing. In *Sphenodon* and some lizards, the inverse has been described. In other species of turtles and lizards, and in crocodilians, two TRTs have been found: intermediate temperatures are masculinizing, whereas lower and higher temperatures are feminizing. Possibly, testing a larger range of temperatures would reveal that species classified in the second pattern actually belong to the third pattern [8].

Within TRT, the pivotal temperature has been operationally defined as that temperature which, in a given species, a population or a clutch, yields 50% males and 50% females [10].

The thermosensitive period (TSP) for gonadal differentiation has been determined in a few species of crocodilians and turtles and in a lizard. For this purpose, shifts from a male- to a female-producing temperature and vice versa were performed at different embryonic stages, and sex ratio was examined at the time of hatching. The duration of TSP represents 18–30% of embryonic development, since the developmental stage of embryos at oviposition is variable. However, in all cases, TSP corresponds to the first stages of gonadal differentiation as revealed by classical histology [8].

The different patterns of TSD and their features (pivotal temperature, transitional range of temperature, thermosensitive period) have been established from incubations at constant temperatures in the laboratory. In nature, nest temperature fluctuates, in particular in shallow nests, which are submitted to nycthemeral rhythm and weather changes [11]. In such conditions, the hatchling sex ratio depends on the proportion of embryonic development that occurs above and below the pivotal temperature during TSP [11, 12].

The reader will find in several previous reviews information on the different patterns of TSD, their features and their ecological and evolutionary implications [1, 2, 13–17]. Other reviews deal mainly with endocrinological aspects of TSD, in particular the role of steroids and steroidogenic enzymes in gonadal sex differentiation [8, 18–23]. Several hypotheses on the molecular mechanisms of TSD have already been formulated [8, 18,

23–27], although molecular approaches are only beginning to be used in reptiles.

In this review, we have selected cellular, molecular and physiological data obtained at the gonadal level only, i.e. data that, in our opinion, are fundamental to elucidate the mechanism of TSD in reptiles and to allow fruitful comparisons with other vertebrates.

Gonadal differentiation and growth as a function of temperature

Gonadal differentiation has been described in several reptiles, including both GSD and TSD species [28]. Some histological descriptions were made long before the discovery of TSD, in species which are now known to exhibit TSD, such as the turtles, *Chrysemys picta* [29] and *Sternotherus odoratus* [30], and the alligator *Alligator mississippiensis* [31]. Recent studies were performed in other TSD turtles, including *T. graeca* [32, 33], *E. orbicularis* [34], *Dermochelys coriacea* [35], *Lepidochelys olivacea* [36, 37] and *Trachemys scripta* [38, 39], and again in the alligator *A. mississippiensis* [24, 40–42]. In the latter, histological studies, immunohistological localization of two structural proteins (laminin and cytokeratin) and ultrastructural studies were carried out [41–43].

Gonadal growth at male- and female-producing temperatures was analyzed in *E. orbicularis* from the beginning of TSP to hatching by determining gonadal protein content [44].

The main morphological changes occurring during gonadal differentiation are similar in all turtle species and in the alligator, although they display some specific characteristics. Figure 1 schematizes these changes from the formation of the gonadal primordium to the structure of the gonads at hatching. It is based mainly on data obtained in *E. orbicularis* and *A. mississippiensis* [34, 41–44].

As in other vertebrates, the gonadal primordia ("genital ridges") develop as thickenings of the coelomic epithelium on the ventromedial surface of the mesonephric kidneys, on each side of the dorsal mesentery. Primordial germ cells are scattered in this epithelium ("germinal epithelium") and, less frequently, in the underlying mesonephric mesenchyme. In turtles, germ cells are in the "posterior germinal crescent" at the end of gastrulation [30, 45, 46]; they then migrate through the dorsal mesentery to reach the gonad anlagen.

During development of genital ridges, epithelial cells are added to the initial mesenchyme in the inner part of the gonads (medulla). These epithelial cells proliferate from different places: the external epithelium of Bowman's capsule of Malpighian corpuscles, the coelomic epithelium bordering mesonephric kidneys on the lateral side of each gonad (at the place of previous nephrostomes) and the germinal epithelium itself.

Cells that proliferate from both the Malpighian corpuscles and the lateral coelomic epithelium are generally small, darkly staining and are organized

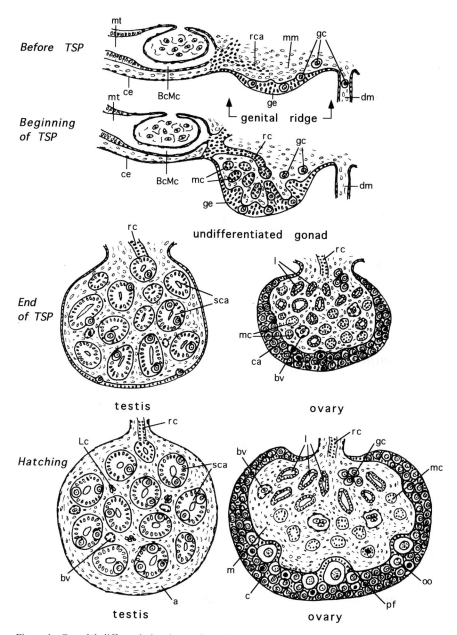

Figure 1. Gonadal differentiation in reptiles with temperature-dependent sex determination: a, albuginea; BcMc, Bowman's capsule of Malpighian corpuscle; bv, blood vessel; c, cortex; ca, cortex anlage; ce, coelomic epithelium; dm, dorsal mesentery; gc, germ cells; ge, germinal epithelium; l, lacunae; Lc, Leydig cells; m, medulla; mc, medullary cords; mm, mesonephric mesenchyme; mt, mesonephric tube; oo, oocyte; pf, primordial follicle; rc, rete cord; rca, rete cord anlage; sca, seminiferous cord anlagen.

into thin cords in the dorsal part of the gonadal primordium. These cords are the anlagen of the rete cords.

Cells that proliferate from the germinal epithelium also give thin cords of epithelial cells which penetrate into the underlying initial mesenchyme. These "medullary cords" or "sex cords" [29] become rapidly outlined with a basement membrane. In *A. mississippiensis*, sex cords are composed of numerous small, irregularly shaped cells, and occasionally of some enlarged somatic cells which have been considered but not determined to be pre-Sertoli cells. The continuity of the basement membrane between the germinal epithelium and the sex cords has been demonstrated by positive laminin immunoreactivity. Cytokeratin is present in the germinal epithelium and in some scattered cells of the medulla (pre-Sertoli cells?). Sex cords and rete cords enter into contact in the inner part of the gonad; thus a mixture of cells from these two types of cords is possible there.

This gonadal structure is observed in both sexes up to the very beginning of TSP. Thus, gonads are usually considered to be undifferentiated at this time; however, sex cords are slightly thinner at female- than at male-producing temperatures. Gonadal sex differentiation occurs during TSP.

In differentiating testes, the germinal epithelium flattens and remains cytokeratin-positive. Germ cells leave it and migrate between the epithelial cells of the medullary cords. In these cords, more and more cells acquire the Sertolian characteristics. Cytokeratin become concentrated in the basal cytoplasm of Sertolian cells. Medullary cords thus form the seminiferous cord anlagen with a rounded contour well delineated by a laminin-immunoreactive basement membrane. During TSP, testis growth is regular due to the development of seminiferous cords in the medulla (Fig. 2).

In differentiating ovaries, the germinal epithelium thickens due to the *in situ* proliferation of epithelial cells and germ cells which give rise to oogonial nests. Thus, an ovarian cortex develops. By the end of TSP, some germ cells have generally entered into meiosis. Strong cytokeratin immunoreactivity is observed in epithelial cells of the cortex, but germ cells are cytokeratin-negative. The cortex is separated from the medulla by a laminin-positive basement membrane. In the medulla, the putative pre-Sertoli cells never become as numerous as in differentiating testes and do not complete differentiation into Sertoli cells. A laminine-positive basement membrane outlines the medullary cords. However, these cords become thin and appear fragmented, being separated by dark interstitial cells and capillary blood vessels. Another difference with seminiferous cords is that in ovarian medullary cords, cytokeratin is localized in the apical cytoplasm of epithelial cells, instead of the basal part in Sertolian cells. In both testes and ovaries, thin rete cords remain localized in the dorsal part of the medulla.

Fragmentation of ovarian medullary cords has often been interpreted as a regression of these cords [36, 37, 39, 41–43]. However medullary cords, or at least some of them, do not completely regress but evolve as small lacunae bordered by a flat epithelium, as clearly shown in the turtles *T. graeca* [33] and *D. coriacea* [35].

During TSP, ovary growth is less important than testis growth. This differential growth can easily be explained by the fact that the ovarian medulla is strongly reduced compared with the testis medulla and that the ovarian cortex has a weak development (Fig. 2). After TSP, a thin albuginea, composed of a few layers of fibroblasts, differentiates and surrounds the testes. At the end of the embryonic development, the diameter of testicular cords decreases somewhat, and consequently the protein content of the testis also decreases (Fig. 2). Leydig cells begin to differentiate in interstitial tissue. Apoptotic cell death is observed in differentiating albuginea and in sparse cells of seminiferous cords (N. Richard-Mercier, M. Dorizzi and C. Pieau, unpublished results). In ovaries, proliferation of germ cells and the entry of these cells into meiosis continue in the cortex which thus strongly thickens. Ovarian protein content also strongly increases and by the end of the embryonic development, it is similar or even surpasses that of testes (Fig. 2).

At hatching, primordial follicles containing growing oocytes are generally observed in the internal part of the ovarian cortex [34]. However in some species, such as the marine turtles *D. coriacea* [35] and *L. olivacea* [36], germ cells do not enter into meiosis during embryonic development.

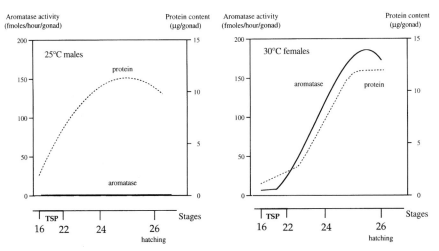

Figure 2. Gonadal aromatase activity and gonadal protein content from the beginning of the thermosensitive period (TSP) to hatching in embryos of *E. orbicularis* incubated at 25°C (male-producing temperature) and at 30°C (female-producing temperature). At 25°C, aromatase activity is very low; protein content increases regularly during and after TSP and somewhat decreases at hatching following the development of testicular cords (future seminiferous cords). At 30°C, aromatase activity increases exponentially during and after TSP and somewhat decreases at hatching; protein content increases weakly during TSP and strongly after TSP, following the development of ovarian cortex. The curves of aromatase activity and protein content are almost parallel (modified from [44]).

These studies of gonadal differentiation in TSD reptiles show that both testicular seminiferous cords and ovarian medullary cords/lacunae derive from proliferation of the surface epithelium (germinal epithelium) of the gonads. Cells that proliferate from the Bowman's capsule of Malpighian corpuscles and the coelomic epithelium on the lateral side of the gonadal primordium give rise to rete cords and could contribute to a limited extent to the edification of testicular and ovarian medullary cords. Follicle cells (future granulosa cells) also appear to derive from the germinal epithelium, whereas interstitial cells between testicular and ovarian medullary cords/ lacunae appear to derive from the initial mesonephric mesenchyme under-lying the germinal epithelium.

Involvement of steroids in gonadal differentiation

At the beginning of the 1970s, when TSD was discovered in turtles, it was well known that embryonic or larval treatments with synthetic estrogens could feminize genetic male individuals up to sex-reversal in birds, reptiles (lizards), amphibians and fish. The development of ovotestes had been obtained in marsupials after early estrogenic treatment of young genetic males in the pouch. However, treatment of gravid females with estrogens or androgens had not modified the sexual differentiation of gonads of the fetuses in eutherian mammals. Androgenic treatments had led to sex rever-sal in some fish and amphibian species but had yielded paradoxical results in other species and in reptiles and birds [18].

In this context, the effects of injection of androgen (testosterone propio-nate) into the eggs were studied in *E. orbicularis* [34] and those of estrogen (estradiol benzoate) were examined in *E. orbicularis* [34] and *T. graeca* [32]. Testosterone did not reverse gonadal sexual phenotype at a female-producing temperature but induced formation of an ovarian-like cortex at the surface of the testes at a male-producing temperature. Estradiol induced various degrees of gonadal feminization at a male-producing temperature, from ovotestis to ovary, depending on the embryonic stage at the time of injection and on the dose of injected steroid. The structural similarity between estradiol-induced and temperature-induced ovaries was striking. Likewise, hormonal-induced ovotestes resembled ovotestes obtained in some individuals at the pivotal temperature in *E. orbicularis* [28, 47], indicating a relationship between gonadal structure, temperature and estrogen levels.

In another approach, histochemical detection of 3β-hydroxysteroid de-hydrogenase-5-ene-4-ene isomerase (3β HSD), a key enzyme in steroid biosynthesis, was carried out, on the same sections, in the adrenal cortex and gonads of *E. orbicularis* and *T. graeca* embryos, using dehydroepi-androsterone (DHA) as a substrate. Before TSP, 3β HSD activity was already strong in the adrenal cortex, and a lower but significant level of enzyme activity was also present in the medullary epithelial cords of the undifferentiated gonad. However, in gonads, activity was higher at masculinizing temperature than at feminizing temperature, whereas in adrenal cortex, activity was similar at both temperatures. During gonadal differentiation (corresponding to TSP), 3β HSD activity increased in testicular cords at a male-producing temperature, whereas it decreased and disappeared in the ovarian medulla at a female-producing temperature. In the adrenal cortex, 3β HSD activity became very strong at both temperatures [34]. These results indicated that steroidogenesis is present very early in both adrenal cortex and gonad anlagen and that in the gonads, but not in the adrenal cortex, synthesis of steroidogenic enzymes or synthesis of some of them is influenced by the incubation temperature.

Based on these preliminary data, two approaches were conducted in *E. orbicularis* and other TSD reptilian species: (i) the comparison of steroid pathways in the gonads of embryos incubated at different temperatures, and (ii) the study of the involvement and role of estrogens and androgens in gonadal sex differentiation.

Evidence of the early presence of active steroidogenic enzymes and temperature-dependent steroidogenesis in the gonads

A preliminary study performed in *E. orbicularis* embryos after TSP had shown that the level of endogenous gonadal steroids was very low and difficult to quantify with the radioimmunoassay techniques used [48]. A sensitive technique, combining two successive chromatographies (HPLC and TLC) and autoradiography was thus developed, allowing both visualization and quantification of the metabolites synthesized by gonads incubated with steroid precursors [49]. In *E. orbicularis*, it is possible to separate gonads from the adrenal-mesonephric complexes as early as the beginning of TSP. Pools of gonads from embryos during or after TSP, and incubated at 25°C (males) or at 30°C (females), were incubated with tritiated pregnenolone, progesterone, dehydroepiandrosterone or androstenedione as substrates. At these different stages, the gonads of *E. orbicularis* were able to metabolize these precursors, showing the presence of active steroidogenic enzymes. A general scheme of steroidogenesis in the gonads was thus deduced from this study (Fig. 3).

Among steroidogenic enzymes, 5α-reductase and 20β-hydroxysteroid oxidoreductase had the highest activity at both temperatures. The 3β HSD activity was significantly higher in differentiating testes at 25°C than in differentiating ovaries at 30°C, thus confirming previous observations made with a histochemical method [34].

Using androstenedione as a substrate, only trace amounts of estrogens were detected on autoradiograms. Therefore, another technique (the tritiated water assay) was used to measure the activity of aromatase, the enzyme complex that converts androstenedione to estrone, and testosterone to

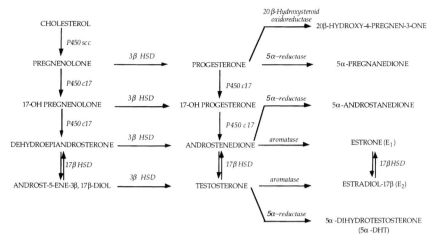

Figure 3. Major steroid pathways in the gonads of *E. orbicularis* embryos. P450scc, cholesterol side-chain cleavage cytochrome P450; P450c17, cytochrome P450 catalyzing both 17α-hydrolase and 17-20-lyase activities; 3β HSD, 3β-hydroxysteroid dehydrogenase-5-ene-4-ene isomerase; 17β HSD, 17β-hydroxysteroid dehydrogenase (modified from [8]).

estradiol-17β (Fig. 3). In the aromatase assay, temperature had no significant effect on enzyme activity. Therefore, the differences observed in gonadal aromatase activity are due to differences in the amount of the enzyme in the gonads. In gonads of embryos incubated at 25°C, aromatase activity remained very low from the beginning of TSP up to hatching. In gonads of embryos incubated at 30°C, aromatase activity was also very low at the beginning of TSP, but increased exponentially during TSP and peaked after TSP; there was a slight decrease of activity around hatching [50]. It is interesting to note that in ovaries, aromatase and growth curves are almost parallel (Fig. 2).

The effects of temperature shifts on gonadal aromatase activity and gonadal structure were subsequently studied. In embryos of *E. orbicularis* shifted from 25 to 35°C (a highly feminizing temperature, [51]) during TSP and then exposed for 1 to 8 days to 35°C, the increase in gonadal aromatase activity was exponential reflecting amplification of aromatase synthesis corresponding to the feminization of the gonad [52].

From these data in *E. orbicularis*, it appears that synthesis (and thus the activity) of at least two gonadal steroidogenic enzymes, 3β HSD and aromatase, is different at male- and female-producing temperatures. Similar results were obtained in some other TSD species. Thus, a higher 3β HSD activity in testes than in ovaries was also shown by steroid metabolism studies in the marine turtle *D. coriacea* (G. Desvages and C. Pieau, unpublished results), and by histochemical detection of the enzyme in *T. graeca* [6] and *A. mississippiensis* [40]. The role of 3β HSD in gonadal sex differentiation remains to be investigated.

Aromatase activity was measured with the tritiated water assay in *D. coriacea* and two crocodilians, *Crocodylus porosus* and *A. mississippiensis*. In *D. coriacea*, the gonadal aromatase activity profiles corresponding to male- and female-producing temperatures, were very similar to those in *E. orbicularis*. Again, shifts from a male- to a female-producing temperature resulted in an increase of aromatase activity only during the thermosensitive period for gonadal differentiation [53].

In crocodilians, the gonads were not separated from the adrenal-mesonephric complexes before hatching. Thus, in embryos, aromatase activity was measured in the gonad/adrenal/mesonephros (GAM) complexes. Activity remained low in the male complex, whereas it increased in the female complex during ovary differentiation. However, differences were found only after TSP: at hatching, high levels of aromatase activity were detected in the ovary of females, whereas in males they were very low and similar in both the testis and the adrenal-mesonephros [54, 55]. Probably, as in *E. orbicularis*, differences in aromatase activity existed between testis and ovary at earlier stages of development, but they could be masked by the aromatase activity of the adrenal-mesonephros part of the GAM. Results suggest that in crocodilians as in turtles, temperature directly or indirectly influences aromatase synthesis in the gonads during embryonic development [55].

Experimental evidence of the involvement of estrogens in ovary differentiation

During TSP, high levels of estrogen content were correlated with high aromatase activities in the gonads of the turtle *D. coriacea* [53]. Moreover, a variety of experiments performed in different TSD species provided evidence for estrogen involvement in ovary differentiation. They consisted in examining the effects of early treatments (before and/or during TSP) with estrogens, antiestrogens or aromatase inhibitors on gonadal differentiation at male- and female-producing temperatures. Estrogens are normally synthesized in the gonads and also in the brain and adipose tissue; antiestrogens compete with estrogens at the level of estrogen receptors and aromatase inhibitors prevent the conversion of androgens to estrogens (see Fig. 3). Among several antiestrogens and aromatase inhibitors used, tamoxifen and two nonsteroidal aromatase inhibitors, fadrozole (CGS 16949 A) and letrozole (CGS 20267) provided the more convincing results.

Let us examine the results obtained in *E. orbicularis* where a complete experimental series was performed (Table 1).

In embryos incubated at 25°C (male-producing temperature), exogenous estrogens (estradiol benzoate or estrone) induced ovarian differentiation. The antiestrogen tamoxifen had previously been reported to have both ago-

Table 1. Effects of early treatments (before and/or during the thermosensitive period) with estrogens (estradiol benzoate or estrone), an antiestrogen (tamoxifen), or nonsteroidal aromatase inhibitors (fadrozole or letrozole) on gonadal differentiation in *E. orbicularis*

Incubation temperature	Treatment	Gonadal structure
25°C (Male producing)	control	testis
	estradiol benzoate or estrone	ovary
	tamoxifen	ovotestis
	tamoxifen + estradiol benzoate	ovotestis
	fadrozole or letrozole	testis
30°C (Female producing)	control	ovary
	estradiol benzoate	ovary
	tamoxifen	ovotestis
	tamoxifen + estradiol benzoate	slightly masculinized ovaries
	fadrozole or letrozole	ovotestis or testis

nistic and antagonistic effects on mammalian tissues. Both effects were obtained at 25°C in *E. orbicularis*. Tamoxifen induced the formation of an ovarian-like cortex at the surface of the testes which thus became ovotestes (agonistic effect). When tamoxifen was applied simultaneously with estradiol benzoate, the gonads were also ovotestes, showing that tamoxifen had prevented the inhibitory action of estrogen on testicular cord development (antagonistic effect) [56]. Fadrozole and letrozole had no apparent effect on testis differentiation [57].

In embryos incubated at 30°C (female-producing temperature), estrogenic treatment somewhat reduced the size of the gonads but did not affect their differentiation into ovaries. Tamoxifen alone led to the development of ovotestes. A slight masculinization of the ovaries was still observed when estradiol benzoate was injected simultaneously with tamoxifen: epithelial cords were present in the medulla, although they were thinner than typical testicular cords [56]. Fadrozole and letrozole induced various degrees of gonadal masculinization, letrozole being more potent than fadrozole. In many cases, ovotestes or testes with typical testicular cords were obtained [57, 58]. In such masculinized gonads, aromatase activity was significantly lower than in ovaries, being close or similar to that in testes of embryos incubated at 25°C [58]. Treatment with letrozole after TSP also resulted in different degrees of gonadal masculinization up to the formation of ovotestes, always exhibiting lower aromatase activity than that of an ovary. Thus, the ovary retains male potential after TSP [59].

Several data, similar to or completing those in *E. orbicularis*, were obtained in other reptilian species with TSD. Estrogenic treatment induced ovary differentiation at a male-producing temperature in all species studied, including crocodilian, lizard and other turtle species [60]. In *C. porosus*, specific binding of tritiated estradiol was found in the gonads of both sexes during the first stages of their sexual differentiation [61]. Accordingly, in the turtle *T. scripta*, the time period during which gonadal differentiation is sensitive to exogenous estrogens was shown to coincide with the thermosensitive period [62]. Moreover, shifts from a male- to a female-producing temperature as well as estrogenic treatments at different stages of this period had similar chronological effects on gonadal structure, the inhibition of testicular cord development preceding the proliferation of germ cells in the cortex [39]. To obtain complete feminization of the gonads, the doses of exogenous estrogen required near the pivotal temperatures were lower than those required below this temperature, showing a synergistic effect of temperature and estrogens on gonadal differentiation [63].

Masculinization of gonads by an antiestrogen at a female-producing temperature was observed only in the turtle *E. orbicularis*. Thus, tamoxifen had an agonistic but not an antagonistic effect in embryos of *A. mississippiensis* [64] and *T. scripta* [65]. However, as in *E. orbicularis*, significant results were obtained with the aromatase inhibitors fadrozole and letrozole in two turtles, *T. scripta* and *C. serpentina*. Fadrozole increased the percentage of males at both pivotal and female-producing temperatures in *T. scripta*, and at a predominantly female-producing temperature in *C. serpentina*. Likewise, in *T. scripta*, letrozole increased the percentage of males at a female-producing temperature [66–68].

Sex-reversed individuals (i.e. females resulting from treatment with an estrogen at a male-producing temperature and males resulting from treatment with an aromatase inhibitor at a female-producing temperature) are viable after hatching and display gonadal endocrine function similar to that of control males and females obtained under the effects of temperature alone. Thus estrogen-induced females of the leopard gecko (*E. macularius*) had normal circulating sex steroids and produced viable offspring as adults [69]. In 8-month-old turtles (*C. serpentina*), the profiles of circulating steroids (testosterone and 17β-estradiol) and the response to treatment with follicle-stimulating hormone were similar in sex-reversed and temperature-induced males and females [70].

Altogether, the results of these experiments show that treatment with estrogens before or during TSP induces ovary differentiation at male-producing temperatures, whereas treatment with antiestrogens or aromatase inhibitors results in ovotestis or testis differentiation at female-producing temperatures. It is therefore clear that estrogens are involved in ovary differentiation. Moreover, it appears that estrogens also play a role in maintaining the ovarian structure. In ovarian differentiation, estrogens act on both the inner part and the surface epithelium of the gonads. In the inner part, they inhibit differentiation of testicular cords from the initial epithelial cords (sex cords), and instead induce differentiation of lacunae bordered by a flat epithelium. At the surface, they stimulate the formation of a cortex from the initial germinal epithelium; cortex development is characterized by the proliferation of germ cells and their entry into meiosis. If estrogens are synthesized at very low levels, testicular cords differentiate, the cortex does not develop and testis differentiation occurs. The differential growth of gonads at a male- and a female-producing temperature follows these estrogenic effects [44].

Possible implication of androgens in gonadal differentiation

As indicated above, exogenous testosterone had no masculinizing effect at a female-producing temperature, but feminized the gonads at a male-producing temperature in *E. orbicularis* [34]. The same "paradoxical" effect was obtained in *T. scripta* [67]. In both cases, it was expected to result from aromatization of testosterone. Aromatase activity is very low in testis. However, aromatization of exogenous testosterone could occur in other tissues, such as brain, which already exhibits *aromatase* transcription before TSP in both sexes [23]. Therefore, the effects of dihydrotestosterone (DHT), a nonaromatizable derivative of testosterone, were studied. Testosterone is converted to 5α-DHT or 5β-DHT through the activity of the 5α-reductase or the 5β-reductase. In *T. scripta* and *A. mississippiensis*, treatment with 5α-DHT had no detectable effect on gonadal differentiation at temperatures yielding 100% males or 100% females [65, 71]. Moreover, in *T. scripta*, 5α-reductase inhibitors did not prevent testis differentiation at a 100% male-producing temperature [67]. These results exclude a major role of androgens compared with that of estrogens [8]. However, when eggs of *T. scripta* were incubated at pivotal temperature, treatment with high dosages of 5α-DHT increased the percentage of males, whereas treatment with 5α-reductase inhibitors increased the percentage of females. Simultaneous administration of 5α-DHT and estradiol led gonads to develop as ovotestes, showing that 5α-DHT prevented the inhibitory effect of estradiol on testicular cord development [67]. A masculinizing effect of 5α-DHT in gonadal differentiation is therefore possible around the pivotal temperature, a condition which often occurs in reptile nests in nature [11].

In *E. orbicularis* embryos, both testes and ovaries produced *in vitro* 5α-pregnancies as the major metabolites of progesterone and 5α-androstanes as the major metabolites of dehydroepiandrosterone and androstenedione (Fig. 3), showing a high and similar 5α-reductase activity in both sexes. However, testes but not ovaries produced 5α-DHT from dehydroepiandrosterone or androstenedione as precursors [49]. This can be simply a consequence of higher aromatase activity in ovaries (see Fig. 3). In *D. coriacea*, both 5α- and 5β-reductase activities in gonads were high and similar at both male- and female-producing temperatures (G. Desvages and C. Pieau, unpublished results). Therefore, contrary to aromatase, the synthesis of

5α- and 5β-reductase in the gonads does not appear to be temperature-dependent in turtles with TSD.

The meaning of gonadal intersexuality

Long before the discovery of TSD, fertile testes with immature oocytes at their surface had been described in turtles captured in nature [72].

Recent data in *E. orbicularis* show that gonadal intersexuality occurs in different conditions of incubation and treatment: at, or close to, the pivotal temperature [28, 47]; at fluctuating temperatures in the laboratory [73] and in nature [11, 74]; under the effects of estrogenic treatment at a male-producing temperature [34, 56]; and under the effects of an antiestrogen or an aromatase inhibitor at a female-producing temperature [56–59].

In hatchlings of *E. orbicularis* incubated at the pivotal temperature (28.5°C), ovaries have a typical structure, but most testes display an ovarian-like cortex with germ cells in meiosis. In some cases, the cortex is almost as developed as that of an ovary, and gonads are ovotestes [28, 47]. At 28.5°C, the future testes and ovotestes become clearly distinct from the future ovaries only by the end of TSP, whereas the differences between a testis differentiating at 25°C and an ovary differentiating at 30°C are already detectable at the beginning of TSP [44].

During TSP, aromatase activity in the future testes and ovotestes at 28.5°C remains low but slightly higher than in testes at 25°C; in the future ovaries at 28.5°C, aromatase activity increases but is generally slightly lower than in ovaries at 30°C (Fig. 4). At the end of embryonic development, aromatase activity in testes and ovotestes at 28.5°C becomes similar to that in 25°C testes, and aromatase activity in 28.5°C ovaries becomes similar to that in 30°C ovaries. During the same period, germ cells begin to degenerate in the cortex of ovaries and ovotestes. Germ cell degeneration continues after hatching. However, not all germ cells degenerate into ovaries, whereas in most cases all cortical germ cells degenerate into ovotestes, which thus become testes. Therefore, in general, the formation of an ovotestis is a transient phenomenon due to a slight increase in estrogen synthesis compared with that in a differentiating testis during TSP; estrogen levels are sufficient to induce the formation of an ovarian-like cortex but not sufficient to inhibit the development of the testicular cords. In addition, at the end of the embryonic life and after hatching, estrogen levels are too low to maintain an ovarian-like cortex at the surface of the gonads. However, in a few individuals, some oocytes may escape degeneration, and several months after hatching, they are found in cortex vestiges at the surface of the testes [44].

Gonadal differentiation at the pivotal temperature in *E. orbicularis*, with transient formation of an ovarian-like cortex at the surface of the male gonad, appears very similar to that described 65 years ago in the turtle

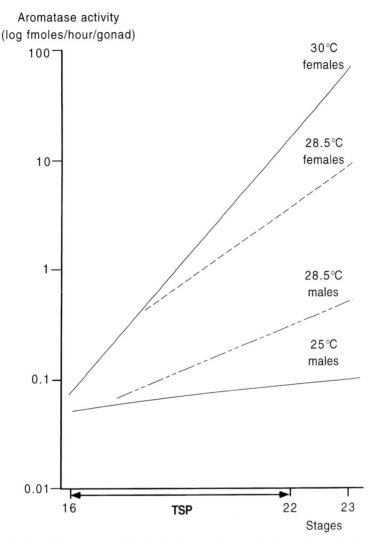

Figure 4. Gonadal aromatase activity during the TSP in *E. orbicularis* embryos incubated at 25°C (male-producing temperature), 28.5°C (pivotal temperature) and 30°C (female-producing temperature). Gonadal aromatase activity in 28.5°C males is slightly higher than in 25°C males; in 28.5°C females it is slightly lower than in 30°C females (modified from [44]).

S. odoratus [30]. The eggs of this turtle were incubated at the ambient laboratory temperature, which probably was close to the pivotal temperature during TSP. Moreover, in this species as in *E. orbicularis*, oocytes at the surface of the testes were found in some individuals after hatching. Observations made in adult turtles show that testes can be functional (i.e. produce spermatozoa), although immature oocytes persist at their surface [72].

As shown above, experiments carried out in *E. orbicularis* with tamoxifen, fadrozole and letrozole at a female-producing temperature led to various degrees of gonadal masculinization [56–59]. The different types of intersexual gonads obtained ranging from a typical ovary to a typical testis had aromatase activity in accordance with their structure [58, 59]. Moreover, they confirmed that the epithelium of testicular cords and that of ovarian lacunae derive from the same lineage (i.e. proliferation of the germinal epithelium, see above). Different steps of transformation of ovarian lacunae into testicular cords or tubes were observed. The epithelium bordering lacunae is flat, whereas that of testicular cords is high (future Sertoli cells). In many cases, lacunae with both a flat epithelium and a high epithelium were found, reflecting subtle regulations between masculinizing and feminizing substances and somewhat different dose effects of these substances in each cell.

Genes involved or that could be involved in gonadal formation and differentiation

Molecular approaches of the mechanism of TSD in reptiles have recently been undertaken in two main directions: (i) cloning and expression of *aromatase* and *estrogen receptor* genes which both appear to play a central role in gonadal differentiation, and (ii) the identification and expression of homologues of genes that have been shown, or expected, to be involved in gonadal formation and/or differentiation in mammals.

Aromatase and estrogen receptor genes

Partial complementary DNA (cDNA) clones or full-length cDNAs of *aromatase* were isolated from two turtles, *Malaclemys terrapin* [75, 76] and *T. scripta* [21] and the alligator *A. mississippiensis* (V.A. Lance, unpublished results). In *M. terrapin*, aromatase transcript levels were measured using a competitive reverse transcription-polymerase chain reaction (RT-PCR) technique, in gonad/adrenal/mesonephros complexes, from the formation of genital ridges to hatching. At a male-producing temperature, *aromatase* transcript levels were below the detection limits of the assay at all stages, whereas at a female-producing temperature they increased exponentially during TSP [23, 77]. As expected, the profiles of *aromatase* expression in the gonad/adrenal/mesonephros complexes of *M. terrapin* parallels the profiles of aromatase activity in gonads of *E. orbicularis* [50]. Therefore, differential aromatase activity as a function of incubation temperature of embryos results from differential regulation of the *aromatase* gene at the transcription level.

Partial cloning of the cDNA and gonadal expression of the estrogen receptor were performed in the turtle *T. scripta* [21, 78]. Estrogen receptor messenger RNAs (mRNAs) were expressed in gonads of putative males and females as early as the beginning of TSP. In differentiating testes, transcripts were found in medullary testicular cords but not in the thin epithelium of the surface. In differentiating ovaries, transcripts were distributed throughout both the cortex and medulla [78]. These observations agree with the results of treatments with estrogens and antiestrogens, showing that estrogens act on both parts, the medulla and the cortex, of the gonad.

Homologues of mammalian genes

Several genes including *WT1*, *SF1*, *SRY*, *SOX9* and *DAX1*, encode transcription factors that are involved in gonadal formation and/or differentiation in mammals [79].

WT1 (Wilms' tumor suppressor 1) and *SF1* (Steroidogenic factor 1, also known as *Ad4BP*, homologue of the *fushi tarazu factor 1*, *FTZ-F1* from the fruit fly) are required for the formation of genital ridges and are thought to be also implicated in gonadal sex differentiation.

SRY (Sex determining region of the Y chromosome) encodes a high mobility group (HMG) box protein that acts as a trigger for testis differentiation. *SOX9* (SRY-related HMG box gene 9) encodes a factor that intervenes downstream of SRY in the differentiation of Sertolian cells; these cells produce the hormone responsible for the regression of Müllerian ducts (anti-Müllerian hormone, AMH) in male embryos.

DAX1 (Dosage Sensitive Sex Reversing -Adrenal Hypoplasia Congenita critical region of the X chromosome) encodes an orphan receptor that is considered to be implicated in ovary differentiation.

Except for *SRY* [3, 80], homologues of these mammalian genes have been characterized in TSD reptiles. Genomic DNA and/or cDNA clones for *A. mississipiensis WT1*, *SF1*, *DAX1* and *SOX9* [81–83] and for *T. scripta WT1* and *SOX9* [84] were sequenced. Moreover, putative cDNA clones for *SF1* and *AMH* were isolated in *T. scripta* [85].

In *A. mississipiensis*, the expression of *WT1*, *SF1*, *SOX9* and *DAX1* was examined in the gonad/adrenal/mesonephros (GAM) complexes before TSP and during most of TSP, or in the gonads alone at the end of TSP and after TSP. *WT1*, *SF1* and *DAX1* expression was analyzed using the RT-PCR technique. The three genes were expressed at both male- and female-producing temperatures at all stages. *WT1* and *DAX1* expression appeared to increase at both temperatures at about the middle of TSP and exhibited no sex difference, whereas *SF1* appeared to be expressed at a lower level in males than in females by the end of TSP [83]. *SOX9* expression was analyzed by two independent methods: RT-PCR and *in situ* hybridization in gonad sections. At two female-producing temperatures, RT-PCR showed only basal levels (likely to be nonfunctional) of *SOX9* expression and *in situ* hybridization failed to detect any expression regardless of the stage. At the male-producing temperature, a basal level of *SOX9* expression was observed by RT-PCR and no *SOX9* transcripts were detected by *in situ* hybridization at earlier stages, i.e., before TSP and during most of TSP. Both techniques showed a significant level of *SOX9* expression in testes by the end of TSP and increasing levels after TSP [82].

In *T. scripta*, the expression of *WT1* and *SOX9* was analyzed on the GAMs of embryos incubated at male- and female-producing temperatures, between the formation of genital ridges and the end of TSP. Northern blot analyses of RNA revealed a single *WT1* transcript and two *SOX9* transcripts in the GAMs, at both temperatures, and at all stages. During TSP, the relative proportion of the two *SOX9* transcripts differed at the two temperatures; whether the differences in the ratio of these transcripts reflects a true effect of temperature was not determined. Analysis of *SOX9* expression was also carried out by *in situ* hybridization of whole mounts. Before and at the beginning of TSP, *SOX9* was expressed in gonads at the two incubation temperatures; by the end of TSP, it was expressed in testes but no longer in ovaries. During this whole period, *SOX9* was expressed in mesonephros of both sexes [84]. Recently, immunohistochemical detection of the SOX9 protein was carried out in another turtle species, *L. olivacea* [86]: results were similar to *SOX9* RNA expression in *T. scripta*. Indeed the SOX9 protein was detected in undifferentiated gonads of embryos at both male- and female-producing temperatures. Then, during TSP, SOX9 expression levels remained positive in differentiating testes, whereas they became negative in differentiating ovaries [86]. Therefore, the regulatory mechanisms of the *SOX9* gene during TSP seem different in turtles and in the alligator.

Differences in the expression patterns of sex determining genes also appear between mammals and TSD reptiles. Thus, in the alligator, *SF1* expression is downregulated during testis differentiation whereas in the mouse, it is downregulated during ovary differentiation. *DAX1* expression is similar for both sexes in the alligator, and not downregulated during testis differentiation as is the case in mouse. Apparently, *SOX9* upregulation in differentiating testis occurs at a later stage in the alligator than in the mouse. As found in the chicken testis [87, 88], the timing of *SOX9* upregulation in the alligator testis seems to coincide with its structural organization, but is not consistent with a role for this gene in the early stages of sex determination [82].

Taken together, these preliminary results show that, except for *SRY*, all the genes encoding transcription factors that are known to be involved in mammalian sex determination have their homologues in TSD reptiles. However, the expression patterns of these genes during gonadal sex differentiation may be different not only between mammals and TSD reptiles but also between different reptile species. As shown above, the *aromatase* gene plays a key role in gonadal differentiation of TSD reptiles.

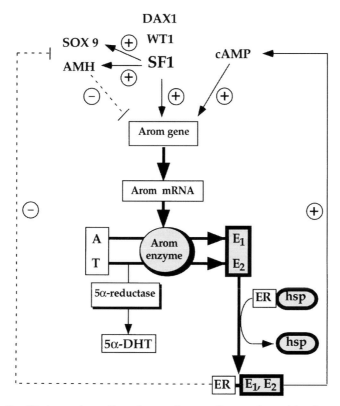

Figure 5. Possible interactions of homologues of mammalian sex-determining factors with the *aromatase* gene in TSD reptiles. $\oplus$, activation of transcription; $\ominus$ repression of transcription; A, androstenedione; E1, estrone; E2, estradiol-17β; ER, estrogen receptor; T, testosterone; DHT, dihydrotestosterone; hsp, heatshock protein.

Hence, several questions can be raised: Do homologues of mammalian sex-determining genes interact directly or indirectly with aromatase, and how? Is transcription of these genes, or of some of them, under estrogen control? How does temperature intervene in these processes? Is the action of temperature similar in species with different TSD patterns?

A scheme presenting possible mechanisms of the action of temperature resulting in the activation or the repression of transcription of the *aromatase* gene was presented previously [8]. Figure 5 includes in this scheme the newly identified factors/genes in TSD reptiles and presents some possible interactions with the *aromatase* gene and the estrogen-estrogen receptor complex. Transcription of the *aromatase* gene is under the control of activating and repressing factors.

As shown in mammals, SF1 appears to act as a transcriptional activator of steroidogenic enzymes, including aromatase [89–91]. Moreover, SF1 appears to also be involved in the activation of expression of the genes

SOX9 and *AMH*, both specific for testis differentiation, with amplification of this transcription when WT1 interacts with SF1. Since, on the other hand, AMH has been shown to repress *aromatase* gene transcription [92], we may consider that in TSD reptiles depending on the incubation temperature, SF1 would act either as a masculinizing factor (via SOX 9 and AMH) or as a feminizing factor (with direct or indirect activation of *aromatase* gene transcription). SOX 9 is expected to be an activator of *AMH* gene transcription in mammals and chicken [93, 94], although it does not seem to be involved in triggering this transcription in chicken [87, 88].

In mammals, DAX1, by interacting with SF1, can prevent the synergistic action of WT1 and SF1 and act as a repressor of *SOX 9* and *AMH* transcription. Thus, by repressing the male pathway, DAX1 can be considered as a feminizing factor [95]. Has DAX1 the same function in TSD reptiles? Is it implicated in the activation of *aromatase* transcription in SF1?

Our model implies that transcription of masculinizing and feminizing genes or of some of them is under the control of estrogens, masculinizing genes being down-regulated, whereas feminizing genes are upregulated by estrogens. In agreement with repression of masculinizing genes by estrogens, a sequence nearly identical to the consensus sequence of estrogen response elements (EREs) has been found in the promoter region of the *AMH* gene in chicken and humans [87, 96]. Repression of *SOX 9* gene transcription by estrogens deserve consideration, since in differentiating ovaries of turtles *SOX 9* transcripts and SOX9 protein disappear [86] when aromatase expression strongly increases [77]. On the other hand, a positive feedback effect of estrogens would easily explain the amplification of aromatase synthesis/activity in differentiating ovaries [8].

At present, there is no evidence for a putative target for temperature. Therefore, the three possible mechanisms of action of temperature described in the previous model [8] can be considered in the new model presented here. Temperature could activate or repress synthesis of either a feminizing factor, a masculinizing factor, or the heatshock proteins (hsps) involved in the binding of estrogens to estrogen receptors; temperature could also be implicated in the dissociation of hsp(s) from the estrogen-estrogen receptor complex which is then activated [8].

Salient points

To date, TSD has been demonstrated in crocodilians, turtles and lizards but not in snakes. During early stages of gonadal development, cells proliferate from the germinal epithelium, and aggregate into epithelial cords in the underlying mesonephric mesenchyme. Then, during the thermosensitive period and depending on the level of endogenous estrogens, these medullary cords yield in testes anlagen of seminiferous cords with high Sertolian epithelium, and in ovaries lacunae with flat epithelium. The estrogen level remains low in testes, whereas it strongly increases and stimulates development of the cortex in ovaries.

Aromatase is therefore a key enzyme in gonadal sex differentiation. Regulation of *aromatase* gene transcription is multifactorial and could be, directly or indirectly, influenced by temperature during the thermosensitive period.

The cDNAs corresponding to *WT1*, *SF1*, *SOX 9*, *DAX1* and *AMH* homologues of mammalian sex-determining genes have been characterized in TSD reptiles. In mammals, SF1 has been demonstrated to activate and AMH to repress *aromatase* gene transcription. Since SF1 also activates *AMH* gene transcription, could this factor not have a pivotal role in *aromatase* gene regulation? The search for other factors that are involved in gonadal steroidogenesis in mammals, such as the steroidogenic acute regulatory (StAR) protein [97] would be of particular interest also in TSD reptiles.

References

1 Bull JJ (1980) Sex determination in reptiles. *Quart Rev Biol* 55: 3−21
2 Bull JJ (1983) Evolution of Sex Determining Mechanisms, Benjamin/Cummings, Menlo Park, CA
3 Coriat AM, Valleley E, Ferguson MWJ and Sharpe PT (1994) Chromosomal and temperature-dependent sex determination: the search for a conserved mechanism. *J Exp Zool* 270: 112−116
4 Charnier M (1966) Action de la température sur la sex-ratio chez l'embryon d'*Agama agama* (*Agamidae*, Lacertilien). *C R Soc Biol* Paris 160: 620−622
5 Pieau C (1971) Sur la proportion sexuelle chez les embryons de deux Chéloniens (*Testudo graeca* L. *et Emys orbicularis* L.) issus d'oeufs incubés artificiellement. *C R Acad Sci Paris* 272D: 3071−3074
6 Pieau C (1972) Effets de la température sur le développement des glandes génitales chez les embryons de deux Chéloniens, *Emys orbicularis* L. et *Testudo graeca* L. *C R Acad Sci Paris* 274D: 719−722
7 Yntema CL (1976) Effects of incubation temperatures on sexual differentiation in the turtle, *Chelydra serpentina. J Morphol* 150: 453−462
8 Pieau C (1996) Temperature variation and sex determination in reptiles. *Bioessays* 18: 19−26
9 Cree A, Thompson MB and Daugherty CH (1995) Tuatara sex determination. *Nature* 375: 543
10 Mrosovsky N and Pieau C (1991) Transitional range of temperature, pivotal temperatures and thermosensitive stages for sex determination in reptiles. *Amphibia-Reptilia* 12: 169−179
11 Pieau C (1982) Modalities of the action of temperature on sexual differentiation in field-developing embryos of the European pond turtle *Emys orbicularis* (*Emydidae*). *J Exp Zool* 220: 353−360
12 Georges A, Limpus C and Stoutjesdijk R (1994) Hatchling sex in the marine turtle *Caretta caretta* is determined by proportion of development at a temperature, not daily duration of exposure. *J Exp Zool* 270: 432−444
13 Janzen FJ and Paukstis GL (1991) Environmental sex determination in reptiles: Ecology, evolution, and experimental design. *Quart Rev Biol* 66: 149−179
14 Ewert MA and Nelson CE (1991) Sex determination in turtles: diverse patterns and some possible adaptive values. *Copeia* 91: 50−69
15 Ewert MA, Jackson DR and Nelson CE (1994) Patterns of temperature-dependent sex determination in turtles. *J Exp Zool* 270: 3−15
16 Viets BE, Ewert MA, Talent LG and Nelson CE (1994) Sex-determining mechanisms in squamate reptiles. *J Exp Zool* 270: 45−56

17 Lang JW and Andrews HV (1994) Temperature-dependent sex determination in croco-
    dilians. *J Exp Zool* 270: 28–44
18 Pieau C, Girondot M, Desvages G, Dorizzi M, Richard-Mercier N and Zaborski P
    (1994) Environmental control of gonadal differentiation. In: The Differences between
    the Sexes, pp 433–448, Short RV and Balaban E (eds), Cambridge University Press, Cam-
    bridge
19 Crews D (1994) Temperature, steroids and sex determination. *J Endocrinol* 142: 1–8
20 Crews D, Bergeron JM, Bull JJ, Flores D, Tousignant A, Skipper JK et al (1994) Tempera-
    ture-dependent sex determination in reptiles: proximate mechanisms, ultimate outcomes
    and practical applications. *Dev Genet* 15: 297–312
21 Crews D (1996) Temperature-dependent sex determination: the interplay of steroid hor-
    mones and temperature. *Zool Sci* 13: 1–13
22 Lance VA (1997) Sex determination in reptiles: an update. *Am Zool* 37: 504–513
23 Jeyasuria P and Place AR (1998) Embryonic brain-gonadal axis in temperature-dependent
    sex determination of reptiles: a role for P450 aromatase (CYP 19). *J Exp Zool* 281: 428–
    449
24 Deeming DC and Ferguson MWJ (1988) Environmental regulation of sex determination in
    reptiles. *Phil Trans R Soc Lond B* 322: 19–39
25 Deeming DC and Ferguson MWJ (1989) The mechanism of temperature dependent sex
    determination in crocodilians: a hypothesis. *Am Zool* 29: 973–985
26 Spotila JR, Spotila LD and Kaufer NF (1994) Molecular mechanisms of TSD in reptiles:
    a search for the magic bullet. *J Exp Zool* 270: 117–127
27 Johnston CM, Barnett M and Sharpe PT (1995) The molecular biology of temperature-
    dependent sex determination. *Phil Trans R Soc Lond B* 350: 297–304
28 Raynaud A and Pieau C (1985) Embryonic development of the genital system. In: Biology
    of the Reptilia, vol. 15, Development B, pp 149–300, Gans C and Billett F (eds), Wiley,
    New York
29 Allen BM (1906) The embryonic development of the rete-cords and sex-cords of *Chrysemys*.
    Am J Anat 5: 79–94
30 Risley PL (1933) Contributions on the development of the reproductive system in the musk
    turtle, *Sternotherus odoratus* (Latreille). *Zeitschr Zellforsch Mikr Anat* 18: 459–543
31 Forbes TR (1940) Studies on the reproductive system of the alligator. IV. Observations
    on the development of the gonad, the adrenal cortex and the Müllerian duct. *Contr Embryol*
    28: 129–156
32 Pieau C (1970) Effets de l'œstradiol sur l'appareil génital de l'embryon de tortue maures-
    que (*Testudo graeca* L). *Arch Anat Micr Morph Exp* 59: 295–318
33 Pieau C (1975) Temperature and sex differentiation in embryos of two chelonians, *Emys
    orbicularis* L. and *Testudo graeca* L. In: Intersexuality in the Animal Kingdom, pp 333–
    339, Reinboth R (ed), Springer, Berlin
34 Pieau C (1974) Différenciation du sexe en fonction de la température chez les embryons
    d'*Emys orbicularis* L (Chélonien): effets des hormones sexuelles. *Ann Embryol Morphog*
    7: 365–394
35 Rimblot F, Fretey J, Mrosovsky N, Lescure J and Pieau C (1985) Sexual differentiation as
    a function of the incubation temperature of eggs in the sea-turtle *Dermochelys coriacea*
    (Vandelli, 1761). *Amphibia-Reptilia* 6: 83–92
36 Merchant-Larios H, Fierro IV and Urruiza BC (1989) Gonadal morphogenesis under con-
    trolled temperature in the sea turtle *Lepidochelys olivacea*. *Herpetol Monogr* 3: 43–61
37 Merchant-Larios H, Ruíz-Ramírez S, Moreno-Mendoza N and Marmolejo-Valencia A
    (1997) Correlation among thermosensitive period, estradiol response and gonadal differen-
    tiation in the sea turtle *Lepidochelys olivacea*. *Gen Comp Endocrinol* 107: 373–385
38 Wibbels T, Bull JJ and Crews D (1991) Chronology and morphology of temperature-depen-
    dent sex determination. *J Exp Zool* 260: 371–381
39 Wibbels T, Gideon P, Bull JJ and Crews D (1993) Estrogen- and temperature-induced
    medullary cord regression during gonadal differentiation in a turtle. *Differentiation* 53:
    149–154
40 Joss JMP (1989) Gonadal development and differentiation in *Alligator mississippiensis* at
    male and female producing incubation temperatures. *J Zool (Lond)* 218: 679–687
41 Smith CA and Joss JMP (1993) Gonadal sex differentiation in *Alligator mississippiensis*, a
    species with temperature-dependent sex determination. *Cell Tissue Res* 273: 149–162

42 Smith CA and Joss JMP (1994) Sertoli cell differentiation and gonadogenesis in *Alligator mississippiensis. J Exp Zool* 270: 57–70

43 Smith CA and Joss JMP (1995) Immunochemical localization of laminin and cytokeratin in embryonic alligator gonads. *Acta Zool* 76: 249–256

44 Pieau C, Dorizzi M, Richard-Mercier N and Desvages G (1998) Sexual differentiation of gonads as a function of temperature in the turtle *Emys orbicularis*: endocrine function, intersexuality and growth. *J Exp Zool* 281: 400–408

45 Pasteels J (1937) Etudes sur la gastrulation des vertébrés méroblastiques. II. Reptiles. *Arch Biol (Liège)* 48: 105–184

46 Cuminge D, Pieau C, Vasse J and Dubois R (1986) Sur l'origine des cellules germinales primordiales chez la Cistude d'Europe (*Emys orbicularis* L.): étude expérimentale des stades gastruléens. *C R Acad Sci Paris* 302 (série III): 557–560

47 Pieau C (1976) Données récentes sur la différentiation sexuelle en fonction de la température chez les embryons d'*Emys orbicularis* L. (Chélonien). *Bull Soc Zool France* 101 (Suppl. 4): 46–53

48 Pieau C, Mignot T-M, Dorizzi M and Guichard A (1982) Gonadal steroid levels in the turtle *Emys orbicularis* L.: a preliminary study in embryos, hatchlings and young as a function of the incubation temperature of the eggs. *Gen Comp Endocrinol* 47: 392–398

49 Desvages G and Pieau C (1991) Steroid metabolism in gonads of turtle embryos as a function of the incubation temperature of eggs. *J Steroid Biochem Mol Biol* 39: 203–213

50 Desvages G and Pieau C (1992) Aromatase activity in gonads of turtle embryos as a function of the incubation temperature of eggs. *J Steroid Biochem Mol Biol* 41: 851–853

51 Pieau C (1978) Effets de températures d'incubation basses et élevées sur la différenciation sexuelle chez des embryons d'*Emys orbicularis* L. (Chélonien). *C R Acad Sci Paris* 286D: 121–124

52 Desvages G and Pieau C (1992) Time required for temperature-induced changes in gonadal aromatase activity and related gonadal structure in turtle embryos. *Differentiation* 52: 13–18

53 Desvages G, Girondot M and Pieau C (1993) Sensitive stages for the effects of temperature on gonadal aromatase activity in embryos of the marine turtle *Dermochelys coriacea. Gen Comp Endocrinol* 92: 54–61

54 Smith CA and Joss JMP (1994) Steroidogenic enzyme activity and ovarian differentiation in the saltwater crocodile, *Crocodylus porosus. Gen Comp Endocrinol* 93: 232–245

55 Smith CA, Elf PK, Lang JW and Joss JMP (1995) Aromatase enzyme activity during gonadal sex differentiation in alligator embryos. *Differentiation* 58: 281–290

56 Dorizzi M, Mignot T-M, Guichard A, Desvages G and Pieau C (1991) Involvement of œstrogens in sexual differentiation of gonads as a function of temperature in turtles. *Differentiation* 47: 9–17

57 Dorizzi M, Richard-Mercier N, Desvages G, Girondot M and Pieau C (1994) Masculinization of gonads by aromatase inhibitors in a turtle with temperature-dependent sex determination. *Differentiation* 58: 1–8

58 Richard-Mercier N, Dorizzi M, Desvages G, Girondot M and Pieau C (1995) Endocrine sex reversal of gonads by the aromatase inhibitor Letrozole (CGS 20267) in *Emys orbicularis*, a turtle with temperature-dependent sex determination. *Gen Comp Endocrinol* 100: 314–326

59 Dorizzi M, Richard-Mercier N and Pieau C (1996) The ovary retains male potential after the thermosensitive period for sex determination in the turtle *Emys orbicularis. Differentiation* 60: 193–201

60 Bull JJ, Gutzke WHN and Crews D (1988) Sex reversal by estradiol in three reptilian orders. *Gen Comp Endocrinol* 70: 425–428

61 Smith CA and Joss JMP (1994) Uptake of ³H-estradiol by embryonic crocodile gonads during the period of sexual differentiation. *J Exp Zool* 270: 219–224

62 Gutzke WHN and Chymiy DB (1988) Sensitive periods during embryogeny for hormonally induced sex determination in turtles. *Gen Comp Endocrinol* 71: 265–267

63 Wibbels T, Bull JJ and Crews D (1991) Synergism between temperature and estradiol: a common pathway in turtle sex determination? *J Exp Zool* 260: 130–134

64 Lance VA and Bogart MH (1991) Tamoxifen sex reverses alligator embryos at male producing temperature, but is an antiestrogen in female hatchlings. *Experientia* 47: 263–267

65 Wibbels T and Crews D (1992) Specificity of steroid hormone-induced sex determination in a turtle. *J Endocrinol* 133: 121–129

66 Wibbels T and Crews D (1994) Putative aromatase inhibitor induces male sex determination in a female unisexual lizard and in a turtle with temperature-dependent sex determination. *J Endocrinol* 141: 295–299

67 Crews D and Bergeron JM (1994) Role of reductase and aromatase in sex determination in the red-eared slider (*Trachemys scripta*), a turtle with temperature-dependent sex determination. *J Endocrinol* 143: 279–289

68 Rhen T and Lang JW (1994) Temperature-dependent sex determination in the snapping turtle: manipulation of the embryonic sex steroid environment. *Gen Comp Endocrinol* 96: 243–254

69 Tousignant A and Crews D (1994) Effect of exogenous estradiol applied at different embryonic stages on sex determination, growth, and mortality in the leopard gecko (*Eublepharis macularius*). *J Exp Zool* 268: 17–21

70 Rhen T, Elf PK, Fivizzani AJ and Lang JW (1996) Sex-reversed and normal turtles display similar sex steroid profiles. *J Exp Zool* 274: 221–226

71 Lance VA and Bogart MH (1994) Studies on sex determination in the american alligator *Alligator mississippiensis*. *J Exp Zool* 270: 79–85

72 Forbes TR (1964) Intersexuality in reptiles. In: Intersexuality in Vertebrates Including Man, pp 273–283, Armstrong CN and Marshall AJ (eds), Academic Press, London

73 Pieau C (1973) Nouvelles données expérimentales concernant les effects de la température sur la différenciation sexuelle chez les embryons de Chéloniens. *C R Acad Sci Paris* 277D: 2789–2792

74 Pieau C (1974) Sur la différenciation sexuelle chez des embryons d'*Emys orbicularis* L. (Chélonien) issus d'oeufs incubés dans le sol au cours de l'été 1973. *Bull Soc Zool France* 99: 363–376

75 Jeyasuria P, Roosenburg WM and Place AR (1994) Role of P-450 aromatase in sex determination of the diamondback terrapin, *Malaclemys terrapin*. *J Exp Zool* 270: 95–111

76 Jeyasuria P, Jagus R, Lance VA and Place AR (1996) The role of P450 arom in sex determination of prototheria and non mammalian vertebrates. In: Molecular Zoology: Advances, Strategies and Protocols, pp 369–386, Ferraris JD and Palumbi SR (eds), Wiley, New York

77 Jeyasuria P and Place AR (1997) Temperature-dependent aromatase expression in the developing diamondback terrapin (*Malaclemys terrapin*) embryos. *J Steroid Biochem Mol Biol* 61: 415–425

78 Bergeron JM, Gahr M, Horan K, Wibbels T and Crews D (1998) Cloning and *in situ* hybridization analysis of estrogen receptor in the developing gonad of the red-eared slider turtle, a species with temperature-dependent sex determination. *Develop Growth Differ* 40: 243–254

79 Swain A and Lovell-Badge R (1999) Mammalian sex determination: a molecular drama. *Genes and Dev* 13: 757–767

80 Spotila LD, Kaufer NF, Theriot E, Ryan KM, Penick D and Spotila JR (1994) Sequence analysis of the ZFY and Sox genes in the turtle, *Chelydra serpentina*. *Mol Philo Evol* 3: 1–9

81 Kent J, Coriat AM, Sharpe PT, Hastie ND and Van Heyningen V (1995) The evolution of WT1 sequence and expression pattern in the vertebrates. *Oncogene* 11: 1781–1792

82 Western PS, Harry JL, Graves JAM and Sinclair AM (1999) Temperature-dependent sex determination: upregulation of *SOX9* expression after commitment to male development. *Dev Dyn* 214: 171–177

83 Western PS, Harry JL, Graves JAM and Sinclair AM (2000) Temperature-dependent sex determination in the American alligator: expression of *SF1*, *WT1* and *DAX1* during gonadogenesis. *Gene* 241: 223–232

84 Spotila LD, Spotila JR and Hall SE (1998) Sequence and expression analysis of WT-1 and Sox9 in the red-eared slider turtle, *Trachemys scripta*. *J Exp Zool* 281: 417–427

85 Wibbels T, Cowan J and LeBoeuf R (1998) Temperature-dependent sex determination in the red-eared slider turtle, *Trachemys scripta*. *J Exp Zool* 281: 409–416

86 Moreno-Mendoza N, Harley VR and Merchant-Larios H (1999) Differential expression of SOX9 in gonads of the sea turtle *Lepidochelys olivacea* at male- or female-promoting temperatures. *J Exp Zool* 284: 705–710

87 Oréal E, Pieau C, Mattei M-G, Josso N, Picard J-Y, Carré-Eusèbe D et al (1998) Early expression of *AMH* in chicken embryonic gonads precedes testicular *SOX9* expression. *Dev Dyn* 212: 522–532

88 Smith CA, Smith MJ and Sinclair AH (1999) Gene expression during gonadogenesis in the chicken embryo. *Gene* 234: 395–402

89 Honda S, Morohashi K, Nomura M, Takeya H, Kitajima M and Omura T (1993) Ad4BP regulating steroidogenic P-450 gene is a member of the steroid and thyroid hormone receptor superfamily. *J Biol Chem* 268: 7494–7502

90 Lala DS, Rice DA and Parker KL (1992) Steroidogenic factor 1, a key regulator of steroidogenic enzyme expression, is the mouse homologue of *Fushi tarazu*-factor 1. *Mol Endocrinol* 6: 1249–1258

91 Lynch JP, Lala DS, Peluso JJ, Luo W, Parker KL and White B (1993) Steroidogenic factor 1, an orphan nuclear receptor, regulates the expression of the rat aromatase gene in gonadal tissues. *Mol Endocrinol* 7: 776–786

92 di Clemente N, Goxe B, Rémy JJ, Cate RL, Josso N, Vigier B et al (1994) Effect of AMH upon aromatase activity and LH receptors of granulosa cells of rat and porcine immature ovaries. *Endocrine* 2: 553–558

93 Morais da Silva S, Hacker A, Harley V, Goodfellow P, Swain A and Lovell-Badge R (1996) Sox9 expression during gonadal development implies a conserved role for the gene in testis differentiation in mammals and birds. *Nature Genet* 14: 62–67

94 Kent J, Wheatley SC, Andrews JE, Sinclair AH and Koopman P (1996) A male-specific role for *SOX9* in vertebrate sex determination. *Development* 122: 2813–2822

95 Nachtigal MW, Hirokawa Y, Enyeart-VanHouten DL, Flanagan JN, Hammer GD and Ingraham HA (1998) Wilms' tumor 1 and Dax-1 modulate the orphan nuclear receptor SF-1 in sex-specific gene expression. *Cell* 93: 445–454

96 Guerrier D, Boussin L, Mader S, Josso N, Kahn A and Picard JY (1990) Expression of the gene for anti-Müllerian hormone. *J Reprod Fertil* 88: 695–706

97 Clark BJ, Soo S-C, Caron KM, Ikeda Y, Parker KL and Stocco DM (1995) Hormonal and developmental regulation of the steroidogenic acute regulatory protein. *Mol Endocrinol* 9: 1346–1355

Genes and Mechanisms in Vertebrate Sex Determination
ed. by G. Scherer and M. Schmid
© 2001 Birkhäuser Verlag Basel/Switzerland

Sex chromosomes, sex-linked genes, and sex determination in the vertebrate class Amphibia

Michael Schmid and Claus Steinlein

Department of Human Genetics, University of Würzburg, Biozentrum, Am Hubland, D-97074 Würzburg, Germany

Summary. In this chapter the different categories of homomorphic and heteromorphic sex chromosomes, types of sex-determining mechanisms, known sex-linked genes, and data about sex-determining genes in the Amphibia have been compiled. Thorough cytogenetic analyses have shown that both XY/XX and ZW/ZZ sex chromosomes exist in the order Anura and Urodela. In some species quite unusual systems of sex determination have evolved (e.g. 0W-females/00-males or the co-existence of XY/XX and ZW/ZZ sex chromosomes within the same species). In the third order of the Amphibia, the Gymnophiona (or Apoda) there is still no information regarding any aspect of sex determination. Whereas most species of Anura and Urodela present undifferentiated, homomorphic sex chromosomes, there is also a considerable number of species in which an increasing structural complexity of the Y and W chromosomes exists. In various cases, the morphological differentiation of the sex chromosomes occurred as a result of quantitative and/or qualitative changes to the repetitive DNA sequences in the constitutive heterochromatin of the Y and W chromosomes. The greater the structural differences between the sex chromosomes, the lesser the extent of pairing in meiosis. No dosage compensation of the sex-linked genes in the somatic cells of the homogametic (XX or ZZ) individuals have been detected. The genes located to date on the amphibian sex chromosomes lead to the conclusion that there is no common ancestral or conserved sex-linkage group. In all amphibians, genetic sex determination (GSD) seems to operate, although environmental factors may influence sex determination and differentiation. Despite the accumulated evidence that GSD is operating in Anura and Urodela, there is little substantial information about how it functions. Although several DNA sequences homologous to the mammalian *ZFY*, *SRY* and *SOX* genes have been detected in the Anura or Urodela, none of these genes is an appropriate candidate to explain sex determination in these vertebrates.

Sex chromosomes in the Amphibia

Homomorphic sex chromosomes

Most amphibian species present morphologically undifferentiated (homomorphic) sex chromosomes [1–4]. This means that in the heterogametic sex, the XY or the ZW sex chromosomes exhibit an identical morphology when studied with the classical cytogenetic techniques (uniform staining of chromosomes). Therefore, the early pioneering studies on amphibian karyotypes failed in the demonstration of differentiated sex chromosomes or yielded contradictory results [5]. Moreover, as no sex-linked genes with their characteristic mode of inheritance were known in the amphibians, other approaches were made to reveal the type of sex-determining mechanisms in these vertebrates. Such experiments were extremely time-con-suming and difficult, but offer a most appealing and convincing method of proof.

The first succesful demonstration of a sex-determining mechanism with non-cytogenetic methods was made by Humphrey [6–8] for the American salamanders *Ambystoma mexicanum* and *Ambystoma trigrinum*. The ex-periment is summarized in Figure 1A. The primordium of the ovary was excised from one side of the body of a genetically female embryo and re-placed by the primordium of a testis taken from a genetically male embryo. Since at this stage of embryogenesis (tail-bud stage) the sex can not be recognized, the appropriate combination (male primordium into female embryo) was achieved in approximately a quarter of all transplantations.

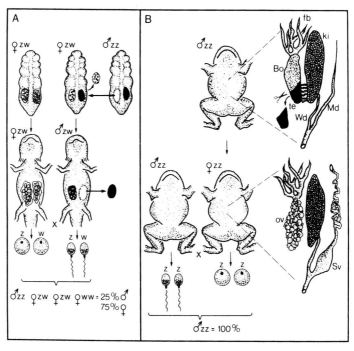

Figure 1. (A) Schematic representation of the sex reversal experiment demonstrating female heterogamety in the Mexican axolotl (*Ambystoma mexicanum*). A detailed explanation is given in the text. The granular tissue in the female embryos represents the primordium of the ovary, and the black tissue represents the primordium of the testis. The cross of a normal female (left) with a sex-reversed female (right) must yield a progeny of 25% males and 75% females if the female sex is heterogametic (ZW). (B) Representation of the sex reversal experiment in the European toad *Bufo bufo* demonstrating male homogamety. Further explanation in the text. The structure of the gonads in the normal and sex-reversed male is shown in the drawings on the right. Bo: Bidder's organ; fb: fat body; ki: kidney; Md: Müllerian duct; ov: ovary; Sv: seminal vesicle; te: testis; Wd: Wolffian duct. The cross of a normal male (left) with a sex-reversed male (right) yields only male progeny if the male is homogametic (ZZ).

During the further development of the embryo, the transplanted testis hormonally modified the contralateral ovary into a functional testis. As soon as this condition was achieved, the transplanted testis was removed. In this way, genetically female embryos were forced to develop into male animals (sex-reversed females) with a testis producing sperm cells from genetically female germ cells. When these sex-reversed females were mated with normal females, the ratio of their male and female progeny was 1:3. This result can only be explained if, as shown in the Figure 1A, the normal *Ambystoma* female is the heterogametic sex (with undifferentiated ZW sex chromosomes) and the male the homogametic sex (ZZ sex chromosomes).

Among the female progeny, two genotypes (ZW and WW), which cannot be differentiated phenotypically, must occur. Indeed, one third of the females (WW individuals) when crossed with normal males (ZZ) produced female progeny exclusively (ZW). The fact that the ZW and WW females are phenotypically identical is a strong indication that the sex chromosomes of this salamander are not only morphologically but also genetically largely homologous. The only genes in which the Z and W chromosomes ought to differ are the opposing sex-determining factors [9].

Sex-reversed males also permit the demonstration of the type of sex-determining mechanism. Such an experiment was performed in the European toad *Bufo bufo*. The technique is shown in Figure 1B. In the sexually mature male toad, the Bidder's organ is located at the anterior part of the testis. This tissue is the incompletely involuted cortex of the embryonic gonad, and has been compared to a rudimentary ovary. Furthermore, the Müllerian duct has been conserved. When the testes of an adult toad were removed, the Bidder's organ developed into a functional ovary and the Müllerian ducts enlarged. When such a sex-reveresed male individual was mated with a normal male, exclusively male progeney ensued. This can be explained if, as represented in Figure 1B, the male *Bufo* is homogametic (ZZ) and the female heterogametic (ZW).

In the African clawed frog, *Xenopus laevis*, sex reversal was induced by treating the developing larvae with estradiol. All tadpoles exposed to estradiol developed as functional females whatever their sex chromosome constitution. Half of these females (ZZ individuals) when crossed with normal males (ZZ) produced male progeny exclusively. In this way, female heterogamety has been demonstrated for *X. laevis* [10–12].

The examples so far described show the occurrence of a ZZ/ZW sex-determining mechanism in amphibians. However, there are also species in which breeding experiments show a mechanism of the XX/XY type. Thus, analyses of the sex ratio in the progeny of sex-reversed females and parthenogenetically bred individuals have demonstrated that the males are heterogametic in the following species of Asiatic frogs: *Rana nigromaculata, R. japonica, R. brevipoda, R. rugosa, Hyla arborea japonica*, and *Bombina orientalis* [11].

Table 1. Sex-determining mechanisms and sex-specific chromosomes found in the Anura

Family	Species	Genetic sex[a]	Sex chromosomes[b]	Morphology[c]	Ref.
Pipidae	*Xenopus laevis*	ZZ/ZW			10–12, 15
Bombina-toridae	*Bombina orientalis*	XX/XY			13
Pelodytidae	*Pelodytes punctatus*	XX/XY			19
Ranidae	*Rana brevipoda*	XX/XY			13
	Rana clamitans	XX/XY			21
	Rana esculenta	XX/XY	XX/XY	X = Y; rpb	22, 23
	Rana japonica	XX/XY	XX/XY	X = Y; ch; rpb	13, 67, 68
	Rana nigromaculata	XX/XY			13
	Rana pipiens	XX/XY			24
	Rana ridibunda	XX/XY			19
	Rana rugosa	XX/XY	XX/XY		13, 62–66
	Rana rugosa	ZZ/ZW	ZZ/ZW		62–66
	Rana temporaria	XX/XY			25
	Pyxicephalus adspersus	ZZ/ZW	ZZ/ZW	Z > W	18, 26, 27
	Pyxicephalus delalandii		ZZ/ZW	Z = W	26
Hylidae	*Hyla arborea japonica*	XX/XY			13
	Gastrotheca riobambae		XX/XY	X < Y	28–31
	Gastrotheca pseustes		XX/XY	X = Y; ch	32
	Gastrotheca walkeri		XX/XY	X > Y	33
	Gastrotheca ovifera		XX/XY	X > Y	33
Centrolenidae	*Centrolenella antisthenesi*		XX/XY	X = Y; ci; ch	34
Rhacophoridae	*Buergeria buergeri*		ZZ/ZW	Z = W; ch; nor	35, 102–104
Bufonidae	*Bufo bufo*	ZZ/ZW			13, 18, 36
Lepto-dactylidae	*Eupsophus migueli*		XX/XY	X = Y; ci	37
	Eleutherodactylus maussi		XX/XY	X = Y; yrt	38
Myobatrach-idae	*Crinia bilingua*		ZZ/ZW	Z < W; ci	39
Leiopelmat-idae	*Leiopelma hochstetteri*		OO/OW		40
	Leiopelma hamiltoni		ZZ/ZW		41

[a] Determined by analysis of the sex ratio in the progeny of sex-reversed animals, partheno-genetically bred individuals, patterns of inheritance and expression of isozymes encoded by the sex chromosomes, or H-Y antigen typing. [b] Determined by cytogenetic analyses. [c] X = Y or Z = W: both sex chromosomes have the same size but differ in their replication banding patterns (rbp), centromeric index (ci), amount of constitutive heterochromatin (ch), presence of a nucleolus organizer region (nor), or the Y is involved in a Robertsonian translocation with an autosome (yrt). X < Y: X is smaller than Y; X > Y: X is larger than Y; Z > W: Z is larger than W; Z < W: Z is smaller than W. In *L. hochstetteri* no Z chromosomes are present.

The cell membrane-associated H-Y antigen or a cross-reactive antigen is found in the gonad of the heterogametic (XY or ZW) sex in all vertebrates so far studied [14–17]. It can be inferred that this antigen is the product of a gene that has been extremely conserved during vertebrate evolution, dating from a common ancestor of fishes, amphibians, reptiles, birds, and mammals. Because H-Y antigen is found in the heterogametic sex, it can function as a useful tool in evaluating the type of sex-determining mechanism (XX/XY or ZZ/ZW) in those amphibian species in which no heteromorphic sex chromosomes have evolved or in which sex reversal experiments have not yet been done. In this way, the heterogametic sex was deduced in *Xenopus laevis* [15], *Bufo bufo, Pyxicephalus adspersus* [18] *Rana ridibunda, Pelodytes punctatus, Pleurodeles waltl, Ambystoma mexicanum* [19], and *Triturus vulgaris* [18].

A summary of the species of Anura and Urodela in which the genetic sex has been assigned by sex-reversal and breeding experiments, H-Y antigen typing or segregation analyses of sex-linked genes, and in which sex chromosomes have been detected by cytogenetic studies, is presented in the Tables 1 and 2 (for review, see also [20]). Although these species still constitute a very small sample, it becomes apparent that in both the primitive and the highly evolved families of the Urodela and Anura, XX/XY and ZZ/ZW sex-determining types coexist. For the third order of the Amphibia, the Apoda (Gymnophiona), neither breeding experiments nor H-Y antigen-typing have been performed, and only few cytogenetic studies have been published.

Heteromorphic sex chromosomes

Initial stages of morphological differentiation between the sex chromosomes

An amphibian species in which the XY chromosomes are still in a very primitive stage of morphological differentiation is the European water frog *Rana esculenta* [22]. In this species male heterogamety was determined early on by breeding experiments [23]. The experimental precondition for the demonstration of the Y chromosome was a technique based on the incorporation of the thymidine analogue bromodeoxyuridine into the replicating DNA during the synthesis (S) phase of the cells (BrdU-replication banding). With this method, the "homomorphic" chromosome pair no. 4 of the male individuals of *R. esculenta* could be identified as XY sex chromosomes. All males have an extremely-late replicating band in the long arm of the Y (Fig. 2a,b,e), which is lacking in the X (Fig. 2c). In this species the sex chromosomes can not be distinguished by other staining or banding techniques (Fig. 2d). This was the first example of vertebrate sex chromosomes in which the heteromorphism could be recognized exclusively by a difference in the DNA replication patterns. The actual cause of the re-

Table 2. Sex-determining mechanisms and sex-specific chromosomes found in the Urodela

Family	Species	Genetic sex[a]	Sex chromo-somes[b]	Morphology[c]	Ref.
Proteidae	*Necturus alabamensis*		XX/XY	X > Y	42
	Necturus beyeri		XX/XY	X > Y	42
	Necturus lewisi		XX/XY	X > Y	42
	Necturus maculosus		XX/XY	X > Y	42, 43
	Necturus punctatus		XX/XY	X > Y	42
Ambysto-matidae	*Ambystoma mexicanum*	ZZ/ZW			6–8, 19
	Ambystoma tigrinum	ZZ/ZW			6–8
Salaman-dridae	*Pleurodeles waltl*	ZZ/ZW	ZZ/ZW	Z = W; lbc	19, 44–48
	Pleurodeles poireti		ZZ/ZW	Z = W; lbc	44–46
	Triturus alpestris	XX/XY	XX/XY	X = Y; ch	49, 50
	Triturus cristatus		XX/XY	X = Y; ch	51
	Triturus helveticus		XX/XY	X = Y; mm	49
	Triturus italicus		XX/XY	X = Y; ch	50
	Triturus marmoratus		XX/XY	X = Y; ch	51
	Triturus vulgaris	XX/XY	XX/XY	X = Y; ch	18, 49, 50
Pletho-dontidae	*Aneides ferreus*		ZZ/ZW	Z = W; ci	52
	Chiropterotriton abscondens		XX/XY	X > Y	2, 53–57
	Chiropterotriton bromeliacea		XX/XY	X > Y	2, 53–57
	Chiropterotriton cuchumatanos		XX/XY	X > Y	2, 53–57
	Chiropterotriton rabbi		XX/XY	X > Y	2, 53–57
	Bolitoglossa subpalmata		XX/XY	X > Y	2, 53–57
	Hydromantes ambrosii		XX/XY	X = Y; ci; ch	58
	Hydromantes flavus		XX/XY	X = Y; ci; ch	58
	Hydromantes imperialis		XX/XY	X = Y; ci; ch	58
	Hydromantes italicus		XX/XY	X = Y; ci; ch	58
	Hydromantes spec. nova		XX/XY	X = Y; ci; ch	58
	Oedipina bonitaensis		XX/XY	X > Y	2, 53–57
	Oedipina poelzi		XX/XY	X > Y	2, 53–57
	Oedipina syndactyla		XX/XY	X > Y	2, 53–57
	Oedipina uniformis		XX/XY	X > Y	2, 53–57
	Thorius pennatulus		XX/XY	X > Y	2, 53–57
	Thorius subitus		XX/XY	X > Y	2, 53–57

[a] Determined by the analysis of the sex ratio in the progeny of sex-reversed animals, partho-genetically bred individuals, patterns of inheritance and expression of isozymes encoded by the sex chromosomes, or H-Y antigen typing. [b] Determined by cytogenetic analyses. [c] X = Y or Z = W: both sex chromosomes have the same size but differ in their centromeric index (ci), amount of constitutive heterochromatin (ch), loop patterns in lampbrush chromosomes (lbc), or pairing arrangement in male meiosis (mm). X > Y: X is larger than Y.

plication asynchrony of the small band in the Y must be a specific DNA sequence. Since repetitive DNA sequences generally begin their replication in the S-phase later than the other regions, it is likely that the late-replicating Y band consists of such repetitive DNA.

Some of the sex chromosomes so far found in Amphibia indicate that heterochromatinization of the Y or W precedes the actual morphological differentiation. The European newts of the genus *Triturus* are characterized

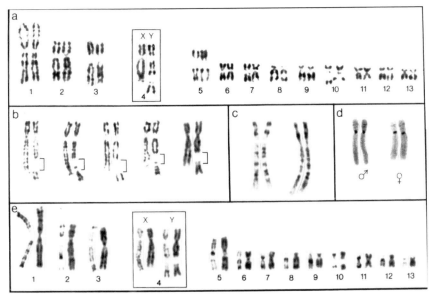

Figure 2. (a) Male karyotype of the European water frog *Rana esculenta* showing the replication banding patterns obtained by BrdU treatment of the cells. The XY chromosome pair no. 4 is framed. Note the late replicating band in the long arm of the Y. Due to the incorporation of BrdU, the condensation of this region is distorted, so that the long arm of the Y appears longer than that in the X. (b) Time sequence of replication bands in the XY pair from early (left) to the very late replication stages (right). Note that in all stages the asynchrony of replication between X and Y is restricted to the late-replicating band in the Y (small brackets). (c) Synchronous replication banding patterns in the XX chromosomes from two females. (d) C-banded sex chromosome pairs no. 4 from a male and a female specimen. (e) The 13 chromosome pairs of *R. esculenta* exhibiting early (left) and late (right) replication banding patterns.

by 24 large meta- to submetacentric chromosomes. After conventional (uniform) staining, the 12 chromosome pairs appear to be homomorphic. The XY sex chromosomes can only be recognized by specific staining of the constitutive heterochromatin. In *Triturus alpestris* and *T. vulgaris* [49] the males have a heteromorphic chromosome pair no. 4. Only one homolog (Y) displays heterochromatic telomeres in the long arms, whereas the telomeres of the other homolog (X) are euchromatic (Fig. 3a–c). The same results have been obtained for the heteromorphic long arms of the XY pair no. 4 of *T. cristatus* and *T. marmoratus* [51]. In the Y there is always distinctly more constitutive heterochromatin than there is in the X. Similar results have been obtained for the heteromorphic long arms of the XY pair no. 2 of *T. italicus* [50]. Of special interest is *T. helveticus* in which no heteromorphism could be demonstrated in the male karyotype (Fig. 3d, 9h). However, in the meiosis of male animals, the long arms of the pair no. 5 show a highly decreased frequency of chiasmata formation (Fig. 9i, k). It is conceivable that in the Y long arm of *T. helveticus* repetitive DNA

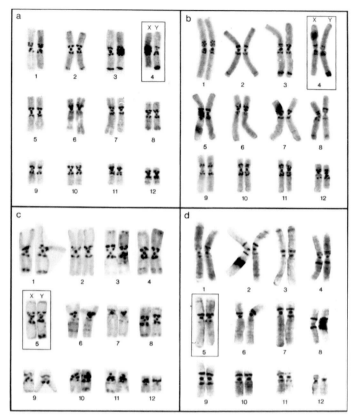

Figure 3. C-banded male karyotypes of some European newts. (a, b) *Triturus alpestris*, (c) *T. vulgaris*, and (d) *T. helveticus* showing the regions of constitutive heterochromatin. The XY chromosomes are framed. In (a, b) T. *alpestris* and (c) *T. vulgaris*, the heteromorphism between X and Y consists of a heterochromatic region located at the telomeres in the long arm of the Y but absent in the X. In (d) *T. helveticus*, no sex-specific heteromorphism of the constitutive heterochromatin can be detected in the sex chromosome pair no. 5; however, the long arms of these chromosomes have a distinctly decreased chiasmata frequency in male meiosis (see Fig. 9h–k).

sequences are already concentrated, although no constitutive heterochromatin can be detected [49].

A slight difference in the morphology of the X and Y was observed in the European salamanders *Hydromantes italicus*, *H. ambrosii*, *H. imperialis*, *H. flavus*, and *H. sp. nova* [58]. In the karyotypes of all females both homologs of the pair no. 14 are acrocentric with identical C-banding pattern. In the males, one homolog (X) is acrocentric and indistiguishable from the female chromosome no. 14, while the other homolog (Y) is submetacentric. The X possesses constitutive heterochromatin close to the centromere in the long arm and at the centromeric region itself, whereas in the Y the heterochromatin is located at the centromeric region and within the short arm.

   In the South American marsupial frog *Gastrotheca pseustes* two different morphs of the Y chromosome were found [32]. In some male individuals, the sex chromosome pair no. 5 (XY$_a$) exhibit the same pattern of constitutive heterochromatin as the XX chromosomes of the females (Fig. 6c, d). In contrast, all the other males from this species show a distinctly heterochromatic telomeric region in the long arm of the Y$_b$ chromosome (Fig. 6a, b). This finding clearly shows that several distinct stages of morphological differentiation can exist within the same population of a single amphibian species.

Advanced stages of morphological differentiation between the sex chromosomes
The first certified highly heteromorphic sex chromosomes in the Anura were discovered in the South African bull frog *Pyxicephalus adspersus* [18, 26, 27]. Male animals have ZZ chromosomes, females the ZW constitution. The W chromosome is considerably smaller than the Z and its short arm is completely heterochromatic (Fig. 4). The same chromosome pair no. 8, which in *P. adspersus* represents the highly heteromorphic ZW pair, is still in an initial stage of morphological differentiation in the closely related *P. delalandii*. Although the chromosomes no. 8 of *P. delalandii* are still of

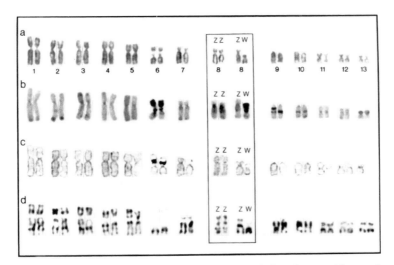

Figure 4. Karyotypes of the South African bull frog *Pyxicephalus adspersus*. The sex chromosome pairs no. 8 are framed. (a) Conventional staining showing homomorphic ZZ chromosomes in the male, and highly heteromorphic ZW chromosomes in the female. (b) C-banding of the constitutive heterochromatin. The W chromosome has a completely heterochromatic short arm. (c) Silver staining showing specific labeling of the nucleolus organizer regions in the short arms of chromosome pair no. 6. (d) Chromosomes with replication banding patterns obtained by BrdU treatment of the cells. Note the synchronous replication patterns in the homologous autosomes and in the ZZ chromosomes of the male and the very late replicating short arm of the W in the female.

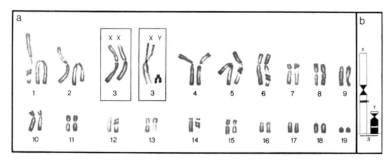

Figure 5. (a) Conventionally stained female karyotype of the North American salamander *Necturus maculosus* showing the homomorphic XX sex chromosome pair no. 3 (framed). The highly heteromorphic XY sex chromosomes from a C-banded metaphase of a male have been included (framed). Note the almost completely heterochromatic Y. (b) Diagrammatic representation of the C-banded XY chromosomes. The arrowhead designates a secondary constriction (Adapted from [43]; with kind permission of Stanley K Sessions).

the same length in the female individuals, they differ from each other by a pericentric inversion and by the amount of heterochromatin [26].

Well-differentiated XY sex chromosomes characterize the males of several species of the American salamanders belonging to the genera *Chiropterotriton*, *Oedipina*, *Thorius*, and *Bolitoglossa* of the family Plethodontidae (Tab. 2). The five species of the American salamander genus *Necturus* have the most highly differentiated XY sex chromosomes yet discovered in the Urodela [42, 43]. These neotenic, perennibranch, and permanently aquatic salamanders with larval morphology belong to the primitive family Proteidae and inhabit the lakes and streams of eastern North America. The genomes of the *Necturus* species are among the largest of the vertebrates (165 pg DNA per diploid cell nucleus in *N. maculosus* [59]). All species (*N. alabamensis*, *N. beyeri*, *N. lewisi*, *N. maculosus*, and *N. punctatus*) have 19 pairs of extremely large chromosomes and highly differentiated XY sex chromosomes (Fig. 5, Tab. 2). The Y chromosomes are about one quarter the size of the X chromosomes and composed almost completely of constitutive heterochromatin [42, 43].

Very distinct heteromorphic XY sex chromosomes were found in the South American marsupial frog *Gastrotheca riobambae* [28–31]. The Y chromosome is the largest element in the karyotype and almost completely heterochromatic (Fig. 6e–h). This is one of the very few vertebrate species having a Y larger than the X. Another peculiarity in the karyotype of *G. riobambae* is the nucleolus organizer region which is located exclusively on the short arm of the X. This leads to a dose ratio of 1:2 in male and female with respect to the number of 18S + 28S ribosomal genes (Fig. 6g, h).

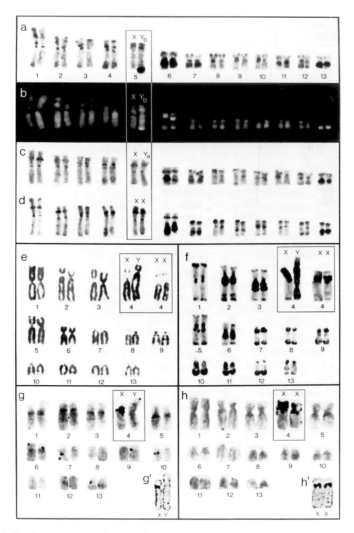

Figure 6. (a–d) Karyotypes of *Gastrotheca pseustes* after C-banding (a, c, d) and quinacrine staining (b). The XX/XY chromosomes are framed. All females (d) possess homomorphic XX chromosomes, whereas among males (a–c) some individuals have inconspicuous Y_a chromosomes (c) and others telomeric heterochromatin in the long arm of the Y_b chromosome (a, b). Note the bright fluorescence of the telomeric heterochromatin in the Y_b shown in (b). (e–h) Karyotypes of *Gastrotheca riobambae* exhibiting conventional staining (e), C-banding (f), *in situ* hybridization with ^{3}H-18S +28S ribosomal DNA (g, h), and silver staining (g', h'). The XX/XY chromosomes are framed. The Y chromosome is the largest chromosome in the karyotype and almost completely heterochromatic. The constriction in the short arm of the X carries the nucleolus organizer.

Exceptional types of sex chromosomes and sex-determining mechanisms
In some species of the Anura very unusual and perhaps unique forms of sex chromosomes have been discovered. These provide new information about the manifold ways in which differentiated Y and W chromosomes have developed in the course of evolution.

A detailed banding analysis on mitotic and meiotic chromosomes of the South American frog *Centrolenella antisthenesi* revealed a new category of XY sex chromosomes [34]. The chromosome pair no. 6 was found to be heteromorphic in the male (Fig. 7a, b). Although both of these chromosomes have the same length, the Y is more submetacentric (arm ratio = 3.5)

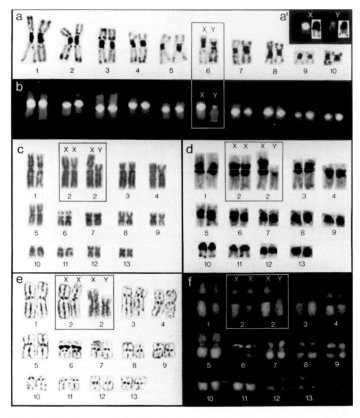

Figure 7. (a, b) Karyotypes of the South American frog *Centrolenella antisthenesi* after C-banding (a) and quinacrine staining (b). The XX/XY sex chromosomes are framed. In (a') the C-banded XY pair has been previously labeled with quinacrine. Although the same amounts of heterochromatin are located in the centromeric and pericentromeric XY regions, only a tiny centromeric Y segment shows bright quinacrine fluorescence. (c–d) Karyotypes of *Gastrotheca walkeri* showing conventional staining (c), C-bands (d), silver staining (e), and mithramycin fluorescence (f). The XX/XY sex chromosomes are framed. Note that extremely little heterochromatin is located on the Y.

than the X (arm ratio = 2.3). In female specimens, the chromosome pair no. 6 is homomorphic and both homologs show the same arm ratio as the single X in the male karyotype. The Y chromosome contains the least amount of AT-enriched heterochromatin, as detected by quinacrine staining (Fig. 7b) or labeling with other AT-specific fluorochromes (DAPI, Hoechst 33258). Only two very small quinacrine-positive bands, one in the centromeric region and the other in the telomeric region of the long arm could be re-cognized in the Y (Fig. 7b). In metaphases prestained with quinacrine and subsequently C-banded, the bright quinacrine bands in the autosomes and in the X are found in exactly the same positions as the dark C-bands. How-ever, in the Y chromosome the centromeric and pericentromeric C-bands are distinctly larger than the quinacrine band (Fig. 7a′). Only a small seg-ment at the centromeric region of the Y is quinacrine positive, whereas its pericentromeric heterochromatin is not labeled by AT-specific fluoro-chromes. These data suggest that the structurally more complex Y chromo-some is the evolutionarily derived condition [34]. Although the X and Y of *Centrolenella antisthenesi* contain fairly similar amounts of repetitive DNA sequences (in terms of C-bands), there are obvious qualitative dif-ferences in the base composition of these sequences.

As in other species of the South American marsupial frogs, the karyo-types of *Gastrotheca walkeri* and *Gastrotheca ovifera* are characterized by differentiated XY sex chromosomes [33]. The Y chromosomes of both species are distinctly smaller than the X chromosomes (Fig. 7c–f). How-ever, whereas the X chromosomes and autosomes contain large amounts of constitutive heterochromatin, extremely little heterochromatin is locat-ed in the Y chromosome (Fig. 7d, f). This is in contrast to all other known amphibian Y chromosomes and the Y chromosomes of most other verte-brates.

One of the most interesting sex-determining mechanisms was discover-ed by Green [40] in the endemic New Zealand frog *Leiopelma hochstetteri*. In this species, a number of unusual cytogenetic phenomena occur. The basic karyotype consists of five pairs of meta- to submetacentric chromo-somes and six pairs of distinctly smaller telocentric chromosomes. Addi-tionally, as first recognized by Stephenson et al. [60], *Leiopelma hochstet-teri* also possesses variable numbers of supernumerary (B) chromosomes in its karyotype (Fig. 8b). These B chromosomes are much smaller than the normal autosomes and not uniform in their morphology. Sex determination in *Leiopelma hochstetteri* was found to be through a supernumerary, uni-valent W chromosome. The females in all populations analyzed invariably have one distinctive supernumerary chromosome not present in males, which can be distinguished from the other supernumerary chromosomes by its distinitve C-band patterns and larger size. Both telocentric and meta-centric isochromosome forms of the W are found in many populations (Fig. 8c). The distribution of constitutive heterochromatin on the W varies between populations, from C-bands restricted to the centromere to almost

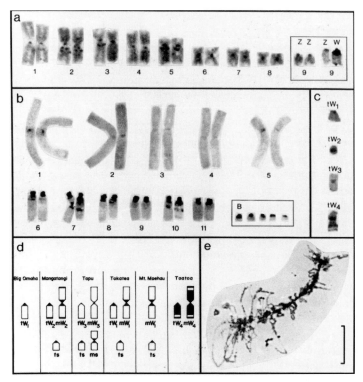

Figure 8. (a) C-banded autosomes and ZZ/ZW chromosome pairs (framed) of *Leiopelma hamiltoni*. The heteromorphism between X and Y is confined to their differing amounts of pericentromeric heterochromatin. (b) C-banded male karyotype of *Leiopelma hochstetteri* with five small supernumerary chromosomes. (c) Four different types of the univalent supernumerary W chromosomes found exclusively in female animals labeled according to the C-band technique. (d) Diagrammatic representation of C-banded W chromosomes and supernumerary chromosomes discovered in several populations in New Zealand. (e) Univalent, supernumerary tW_1 chromosome in lampbrush condition from a 1.0 mm oocyte. Numerous lateral loops indicate transcriptional activity. Bar = 10 μm (Adapted from [40, 41]; with kind permission of David M. Green).

completely heterochromatic W chromosomes (Fig. 8d). In lampbrush preparations, the W chromosomes are always univalent not associating or synapsing with any other chromosome and exhibiting a characteristic lampbrush morphology (Fig. 8e). These data indicate conclusively that the large telocentric or metacentric W chromosomes are morphological variants of a W chromosome in an 0W (female) – 00 (male) system of sex determination. There is neither evidence that the W is a fragment of another chromosome nor that it is a member of a multiple ZW_1W_2/ZZ system. Because fixed chromosomal sex-determining systems that consists solely of a W (or Y) without a Z (or X) have not yet been reported [61], *Leiopelma hochstetteri* is unique among vertebrates.

Geographical variability of sex chromosomes

Some of the heteromorphic sex chromosomes found in the Amphibia can be restricted to isolated populations. The plethodontid salamander *Aneides ferreus* is found along the North American west coast, from northern California to northern Oregon, with an isolated population on Vancouver Island in British Columbia. Extensive cytogenetic studies showed that in the populations from Vancouver Island and northern California two different modes of ZZ/ZW sex chromosomes exist [52]. In the Vancouver Island population, all females are heterozygous for a telocentric and a metacentric chromosome no. 13 (ZW chromosomes), whereas males are homozygous for a telocentric chromosome no. 13 (ZZ chromosomes). In northern California, all males are homozygous for the telocentric chromosome no. 13, but the females may be either identical to males or heterozygous for a telocentric and a metacentric chromosome no. 13. Finally, the *Aneides ferreus* population in Oregon is polymorphic for a pericentric inversion in the chromosome pair no. 13, which is either telocentric or acrocentric. This population is in Hardy-Weinberg equilibrium with respect to the three alternative forms of the chromosome pair no. 13, and sex-specific chromosomes were not found.

   The most intriguing example of an explosive geographical and morphological variability of sex chromosomes, as well as a shift from one sex-determining mechanism to the other, was unraveled by extremely thorough and high-quality cytogenetic analyses in the frog *Rana rugosa* by our Japanese colleagues [13, 62–66]. Seven populations belonging to the northern subgroup of the eastern group, including the Asahikawa and Sapporo populations in the Hokkaido region, the Hirosaki, Akita and Inawashiro populations in the Tohoku region, the Murakami and Kanazawa populations in the Hokuriku region and the Katata population in the Kinki region in the southern subgroup, have a chromosome pair no. 7 which represents ZW sex chromosomes. The Z is acro- or submetacentric, while the W is metacentric. By C-banding and replication banding the different forms of the Z were classified as Z^A, Z^B, Z^C, Z^D and Z^0, and the different forms of the W as W^1 and W^2. Five populations of the southern subgroup of the eastern group, including the Toba population in the Kinki region, as well as the Oigawa, Hamakita, Miyakoda and Yonezu populations in the Chubu region, have a chromosome pair no. 7 which are XY sex chromosomes. The X is metacentric and the Y is acro- or submetacentric. By C-banding and replication banding the two forms of the Y were classified as Y^A and Y^B, while the X was similar to the W^1 of the northern subgroup. Seven populations of the intermediate subgroup of the eastern group, including the Daigo, Hitachiota, Ashikaga, Maebashi, Machida, Kamogawa and Isehara populations in the Kanto region show no heteromorphic sex chromosomes in males or females. The sex-determining mechanisms of these populations is still unclear. The pair no. 7 consists of homomorphic acrocentric chromosomes (7^C) in both the males and females. Finally, also four populations in the western group, including the Okayama, Kumano and Gotsu populations in the Chugoku region and the Nagayo population in the Kyushu region, have no heteromorphic sex chromosomes in males or females. However, this group evidently is characterized by homomorphic XY sex chromosomes on the basis of the results obtained by breeding experiments using sex-reversed animals. The pair no. 7 consists of submeta- or acrocentric chromosomes (7^A, 7^B) in males and females. Thus, at the beginning of the invasion of *Rana rugosa* into Japan, the chromosome pair no. 7 apparently evolved into different types of sex chromosomes within the various geographical regions.

Sex chromosomes in meiosis

The pairing arrangement of chromosomes in prometaphase and metaphase of the first meiotic division can yield important information about the presence of sex chromosomes. This is especially necessary in those species where demonstration of heteromorphic sex chromosomes in mitosis is negative. The more structural differences there are between the X and the Y or the Z and W chromosomes, the smaller is their pairing segment during meiosis. In *Triturus* [49, 51] the crossing-over between the euchromatic long arm of the X and the long arm of the Y with its telomerically located heterochromatin is reduced or even completely supressed (Fig. 9a–f). The genetic homology between the short arms of these XY sex chromosomes has not been influenced by the differentiation of the long arms. Therefore, the short arms pair in the same way as the autosomes. In *T. helveticus*, no sex-specific heteromorphism of the heterochromatin patterns have been found [49]. However, the long arms of the chromosomes no. 5 have a greatly reduced crossover frequency in male meiosis (Fig. 9h–k). It was concluded that these are the XY sex chromosomes. There already may be a structural change in the long arm of the Y that is not yet detectable cytologically. The reduced crossover frequency between the X and Y long arms in *Triturus* promotes further morphological differentiation of these sex chromosomes.

In male diakinesis of *Hydromantes*, the advanced XY sex chromosomes are usually joined by only one chiasma, which occurs in a terminal or an intercalary region of the long arms. Chiasmata within the heteromorphic short arms or in the pericentromeric regions of the long arms have not been observed [58].

In *Gastrotheca riobambae*, homology between the highly heteromorphic XY sex chromosomes is almost completely lost. In diakinesis, a long, stretched bivalent can be observed, as is the case in the meiosis of mammals. This sex bivalent consists of terminally connected XY chromosomes. A sex vesicle is not formed during the male meiotic prophase of *Gastrotheca riobambae* [28, 29].

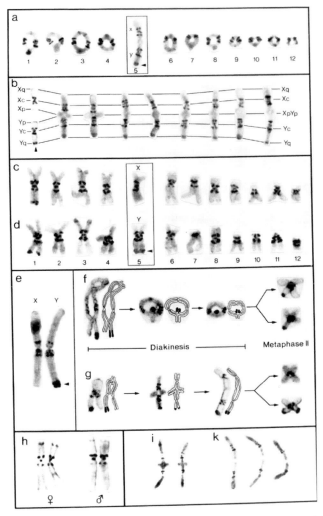

Figure 9. C-banded meiotic chromosomes of male newts of the genus *Triturus*. (a) Diakinetic chromosomes of *Triturus vulgaris*. The homomorphic short arms of X and Y form chiasmata, whereas the heteromorphic long arms remain unpaired and point to opposite directions. The autosomes form bivalents with chiasmata in both arms. (b) Selected examples of XY bivalents of *T. vulgaris*. The same chromosome regions are connected by lines. At the left-hand side the mitotic X and Y chromosomes are depicted. (c, d) X- and Y-carrying metaphases II of *T. vulgaris*. (e–g) Mitotic XY chromosomes (e) and successive stages of diakinetic bivalents of *Triturus alpestris*. (f) Chiasmata in both short and long XY arms lead after their terminalization to ring like sex bivalents and to heteromorphic telomeres in the chromatids of the long arm in metaphase II. (g) Chiasmata located exclusively in the XY short arms lead to end-to-end associated bivalents with diametrically opposed long arms; consequently, in metaphase II the telomeres of the long arms have to be homomorphic. (h) Mitotic no. 5 chromosomes of female and male *Triturus helveticus*. (i, k) Diakinetic no. 5 bivalents of *Triturus helveticus* showing their characteristic end-to-end pairing between the short arms.

In female meiosis, lampbrush chromosomes with their manifold and complex structures permit a very accurate morphological comparison between homologous chromosomes. In the salamanders *Pleurodeles waltl* and *Pleurodeles poireti*, a heteromorphic region was identified in the lampbrush bivalent no. 4, indicating female heterogamety [44–46, 69]. In these species, no heteromorphism could be detected in the mitotic metaphase chromosomes no. 4 by banding techniques [70, 71].

Evolution of sex chromosomes

The evolution of heteromorphic sex chromosomes from one originally homomorphic chromosome pair was probably not the result of a single structural change, but most likely involved several subsequent steps [9]. Because evolutionary processes cannot be reproduced experimentally, the individual changes taking place over the course of chromosome evolution can only be reconstructed by means of comparative studies.

Several of the initial stages of sex chromosome differentiation have been conserved in primitive vertebrates. Thus, it became possible to create evolutionary series in which the increasing structural complexity of the sex chromosomes could be reconstructed. Beçak et al. [72] recognized an evolutionary series of this kind in the karyotypes of snakes, which possess W chromosomes in differing stages of morphological differentiation. In the snakes of the primitive family Boidae, all chromosome pairs of both males and females are still homomorphic. In the specialized family Colubridae, there are many species in which the females exhibit heteromorphic ZW sex chromosomes. Although the W chromosomes are still of the same length as the Z chromosomes, they differ from them by a pericentric inversion. Finally, in the highly evolved families Crotalidae and Viperidae, the W chromosomes are reduced to very small heterochromatic elements similar to the W chromosomes of birds. According to Ohno [9], the primary step in the differentiation of ZW or XY chromosomes is an isolation mechanism that partly or completely prevents free meiotic crossing-over between these chromosomes. An inversion, as found in the W chromosomes of the colubrid snakes is thought to be the cause of such an isolation mechanism [72]. The isolation would provide the precondition for the differential accumulation of opposing sex-determining genes on the originally homologous chromosomes. The pericentric inversion *per se* does not inhibit crossovers, but it creates a situation in which the lack of crossing-over is of great selective advantage. If a crossover occurs within the inverted segment, gametes with partially deleted and duplicated chromosomes are produced. Following the development of the pericentric inversion on the W and Y chromosome, heterochromatinization of these chromosomes starts, whereas the structure of the Z and X chromosomes is largely preserved. Finally, the W and Y chromosomes are reduced to small heterochromatic elements by successive deletions [9].

Studies on constitutive heterochromatin and a repetitive satellite DNA associated with the W chromosomes of snakes indicate that the primary step in the differentiation of the W chromosome was not a pericentric inversion but the development of specific repetitive DNA sequences (heterochromatinization) in the W [73–76]. Thus, the analytical density gradient centrifugation of DNA and *in situ* hybridization on metaphase chromosomes revealed a specific satellite DNA located on the highly differentiated W chromosomes of the advanced snake families. The same W-associated DNA was also demonstrated in those highly evolved snake species that still possess undifferentiated (homomorphic) ZW chromosomes. Therefore, pericentric inversions and deletions seem to be the result, and not the cause, of sex chromosome differentiation [73].

A number of the known sex chromosomes of amphibians support the assertion that one of the initial steps in the evolution of sex chromosomes was an accumulation of repetitive DNA in the W and Y chromosomes. Thus, in the primitive Y chromosomes of *Triturus* and *Gastrotheca pseustes*, the only visible difference between the X and Y is a very small heterochromatic band in the Y. The same is the case in the primitive W of *Leiopelma hamiltoni* (Fig. 8a). Of special interest is the even more primitive Y of *Rana esculenta*, which differs from the X by only a single late-replicating euchromatic band. Singh et al. [73] have proposed that the sex chromosome-associated satellite DNA in the primitive, undifferentiated W chromosome of snakes could bring about asynchrony in the DNA replication pattern of Z and W, thus reducing the frequency of crossing-over between them, which is the prerequisite for further differentiation of the sex chromosomes. Therefore, it is conceivable that repetitive DNA sequences are enriched in the late-replicating Y band of *Rana esculenta*.

In the more advanced Y and W chromosomes of the Amphibia, inversions are already present, as shown by *Hydromantes*, *Aneides*, *Pyxicephalus delalandii*, and *Eupsophus migueli*. Finally, most of the highly evolved Y and W chromosomes are reduced to small, almost completely heterochromatic elements, as in *Necturus*, *Pyxicephalus adspersus*, and certain Neotropical plethodontid salamanders. These are very similar to the Y and W chromosomes in mammals and birds. Y- and W-linked genes, which were once alleles of X- and Z-linked genes, were lost in the course of heterochromatinization. Because this process was very slow, a regulatory system compensating for the monosomy of the X- and Z-linked genes in the heterogametic sex could have evolved simultaneously. However, those genes determining the development of the heterogametic sex could not be permitted to be affected by the heterochromatinization. Nothing can yet be said about the location of these sex-determining genes within the Y and W chromosomes. It is possible that they are located in the few, small euchromatic regions still preserved in these chromosomes.

The exceptional category of Y chromosomes detected in *Gastrotheca walkeri* and *Gastrotheca ovifera* does not fit the evolutionary model of early heterochromatinization of the Y and the W chromosomes mentioned above. However, it is possible that sex-specific repetitive DNA sequences that escape demonstration with C-banding have accumulated in these differentiated Y chromosomes. With regard to the 0W/00 system found in *Leiopelma hochstetteri*, Green [40] proposed that it originated from a primordial ZW/ZZ type through loss of the Z chromosome. Evidence for this assumption is found in the related species *Leiopelma hamiltoni*, in which the ZZ/ZW system operates [41, 77].

Sex-linked genes in the Amphibia

Absence of dosage compensation of sex-linked genes

In amphibian species with recognizable heteromorphic sex chromosomes, there is no indication of a dosage compensation mechanism in the homogametic (XX or ZZ) sex. Typical sex-chromatin bodies have not yet been found in the XX or ZZ nuclei of somatic tissues in species with sex chromosomes in an initial stage of differentiation, or in species with highly heteromorphic sex chromosomes [22, 28–30, 49]. For instance, in neither male *Pyxicephalus adspersus* (ZZ), nor female *Gastrotheca riobambae* (XX) does any chromosome show a tendency towards positive heteropycnosis in the mitotic prophase of somatic cells.

Treatment of cells with BrdU in the S-phase of the cell cycle allowed the analysis of the replication banding patterns in the chromosomes of female (XX) *Rana esculenta* and male (ZZ) *Pyxicephalus adspersus* [22]. Both of the X or Z chromosomes replicated synchronously (Fig. 2c, 4d), and there was no indication of delayed replication as in the late-replicating, inactivated X chromosome of female mammals. The same was found in the BrdU-banded metaphases of female (XX) *Gastrotheca riobambae* [31]. These cytogenetic data provide strong evidence that in amphibians, as in male (ZZ) birds [9, 78–80], dosage compensation of the X- or Z-linked genes does not exist in the homogametic sex.

Linkage analyses and gene mapping on sex chromosomes

Various problems exist in the studies on genetic linkage and physical gene mapping in the amphibians. There are only a few laboratory stocks of known genotype at multiple loci. Most amphibians have a long generation time. Therefore, the crossing of stocks that differ in two or more loci to produce F_1 hybrids and test crossing for linkage are time consuming. With the exception of *Xenopus laevis* there are only few gene probes available for physical mapping via fluorescence in situ hybridization (FISH). Furthermore, the induction of multiple bands along the chromosomes, which is the precondition for an accurate physical gene mapping, is not possible in the Amphibia using the conventional protocols of G-, R-, or Q-banding. The latter can only be accomplished with the more sophisticated BrdU/dT banding technique [120]. FISH using biotinylated gene probes on BrdU/ dT-banded chromosomes of *Xenopus laevis* kidney cell cultures has just recently been initiated with success in our laboratory [101].

The first known example of sex linkage of an enzyme phenotype was in the salamander *Pleurodeles waltl* [47, 48]. In this species, both alleles of a sex-linked peptidase gene are expressed in the erythrocytes. The peptidase is a dimeric enzyme and a hybrid enzyme can be detected in heterozygotes. Sex linkage of the *peptidase-1* gene was also demonstrated for the closely related *P. poireti* [69]. Unfortunately, no other genes have been localized on the sex chromosomes of the Urodela.

More is known about the location of genes on the sex chromosomes of the anuran genus *Rana*. By extensive electrophoretic studies Elinson [21], Wright and Richards [24], and Wright et al. [81] mapped several genes on the sex chromosomes of these frogs. An example of a sex-linked gene, *aconitase-1*, in *Rana clamitans* is illustrated in Figure 10. When males heterozygous for *aconitase-1* were crossed with homozygous females, all the F₁ males were homozygous and all the F₁ females were heterozygous. When both male and female were heterozygous for the same *aconitase-1* alleles, 27% of the F₁ were males homozygous for one allele, 27% were females homozygous for the other allele, and 46% were heterozygous males and females. Finally, in crosses between a heterozygous female and homozygous male, the F₁ showed no relationship between *aconitase-1* phenotype and sex [21]. These results clearly show that the *aconitase-1* gene is located in the homomorphic sex chromosomes of *R. clamitans* and that the male is the heterogametic (XY) sex.

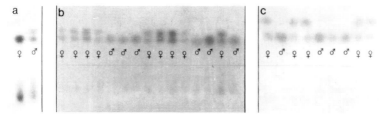

Figure 10. Starch gel plates stained for aconitase-1 isozymes in the North American frog *Rana clamitans*. In (a), the patterns of the homozygous female crossed to a heterozygous male are shown. (b) and (c) demonstrate the patterns of 24 of their offsprings. Note that all heterozygous offspring are female, whereas all homozygous offspring are male. This pattern is consistent with the assumption that in the parents the aconitase-1 alleles c and a were linked to the sex chromosomes as follows: female X^c X^c and male X^a Y^c. Gels in (a) and (b) were run in EDTA-borate-Tris buffer, the gel in (c) in Tris-citrate buffer (Adapted from [21]; with kind permission of Richard P. Ellinson).

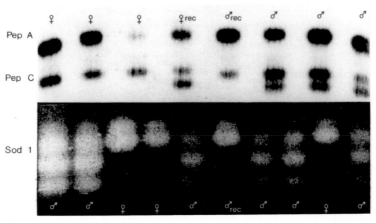

Figure 11. Starch gel plates stained for peptidase isozymes (above) and superoxide dismutase-1 isozymes (below) in the North American frog *Rana pipiens*. Pep-C shows either a single band (alleles *b/b*) or a double band (alleles *b/c*). The double-band pattern is typical for a monomeric enzyme. The samples are from the offsprings of a homozygous X^bX^b female and a heterozygous X^bY^c male. The female offspring are predominantly of the b/b type, whereas the male offsprings are predominantly of the *b/c* type. Two rare recombinants are indicated by "rec". The Pep-A is invariant in this cross. SOD-1 shows either three-band (*b/c*) or single-band (*b/b*) patterns typical for a dimeric enzyme structure. The samples are from offspring of a homozygous X^bX^b female and a heterozygous X^bY^c male. All of the female offspring are X^bX^b homozygous and all but one of the male offspring are X^bY^c heterozygous. The single male X^bY^b recombinant is indicated by "rec" (Adapted from [24]; with kind permission of David A. Wright).

In *Rana pipiens*, sex linkage was demonstrated for peptidase-C (*Pep-C*) and superoxide dismutase-1 (*SOD-1*) [24]. Figure 11 shows that the F_1 offspring of a heterozygous male for *Pep-C* crossed with a homozygous female are mostly of the two parental types, heterozygous males and homozygous females. Recombinant males and females are rare. This mode of inheritance was not found when *Pep-C* heterozygous females were crossed with homozygous males. The analysis of the F_1 offspring of males heterozygous for *SOD-1* shows a mode of inheritance similar to that of *Pep-C* (Fig. 11) [24]. These results demonstrate that the male is heterogametic in *R. pipiens*. Furthermore, the recombination frequencies between the sex-determining gene and the *Pep-C* and *SOD-1* loci, as well as between each of the *Pep-C* and *SOD-1* loci, allowed construction of a linkage map. The sex-determining gene is more distant from *Pep-C* (12.1% recombinants) than it is from *SOD-1* (8.6% recombinants).

There are several closely related species within the *Rana pipiens* complex (*R. pipiens*, *R. sphenocephala*, *R. palustris*, *R. blairi*, and *R. berlandieri*) that form viable hybrids. These frogs show a large number of species-specific allelic differences at a large number of loci. Experiments using such *Rana* hybrids and the analyses of backcross hybrid progeny [81, 82] have provided evidence for sex linkage of the genes for phosphoglucomutase-1, fructose-1,6-diphosphatase, alcohol dehydrogenase-2, α-gluco-sidase, lactate dehydrogenase-B, and mannose phosphate isomerase. The most striking feature is that the genes for sex determination are located on different chromosomes in the northern species (*R. pipiens*, *R. palustris*, and *R. blairi*) than in the southern species (*R. sphenocephala*, *R. berlandieri*), although their karyotypes are nearly identical [81]. The same results were obtained for *Rana japonica* [83], *R. nigromaculata*, and *R. brevipoda* [84].

Graf [85, 86] reported that in the clawed frog *Xenopus laevis* the sex-determining locus is linked with the mitochondrial malic enzyme (mME), and that the recombination rate between them is 6.1%. The recombination frequencies between *mME* and *GPD-1*, and between *GPD-1* and *SOD-1* are 23% each.

It appears that there is no common ancestral or conserved sex linkage group in the Amphibia. Thus, the peptidase gene which is sex-linked in *Pleurodeles waltl* and *P. poireti*, is not homologous with the *Pep-C* gene in *Rana pipiens*. In *Rana catesbeiana*, the gene coding for LDH-B is sex-linked, but in *Rana clamitans* it is located on an autosome.

Repetitive DNA sequences on sex chromosomes of Amphibia

The chromosomal regions classified as C-bands are highly enriched in repetitive DNA sequences [87]. The Y and W chromosomes of many amphibian species are either entirely or partially C-band positive and therefore probably enriched in repetitive DNA sequences. In reptiles, comparative analyses of the DNAs of male and female snakes have revealed the presence of a particular satellite DNA (Bkm) specific to the ZW females [88]. This satellite DNA is enriched in simple GATA/GACA sequences [89], and the number of GATA/GACA sequences parallels the degree of hetero-chromatinization of the W chromosomes in snakes [90].

Synthetic oligonucleotide probes specific for various simple DNA repeats have been used to investigate amphibian sex chromosomes [4]. The $(GATA)_4$ sequence which is the major repeat of Bkm, could not demonstrate a sex-specific signal at the high molecular weight range. In the apodan species *Ichthyophis* and the urodelan species *Necturus* this sequence is underrepresented (Fig. 12). In the anuran species *Gastrotheca riobambae*, where the X and Y sex chromosomes are highly heteromorphic and the Y is loaded with constitutive heterochromatin, GATA sequences are not very abundant. On the contrary, another simple DNA sequence, $(CA)_n/(GT)_n$ is ubiquitously interspersed throughout the genome. With $(GACA)_4$ apparent sex-specific differences could not be marked. However, in *Ichthyophis* and *Necturus* a sex-specific signal can be detected among the male individuals (Fig. 12). A distinct sex-specific signal could be obtained in the female (ZW) of the bull frog *Pyxicephalus adspersus* with $(GAA)_6$ (Fig. 12), and in the male (XY) of the newt *Triturus cristatus carnifex* with $(TCC)_5$ [4].

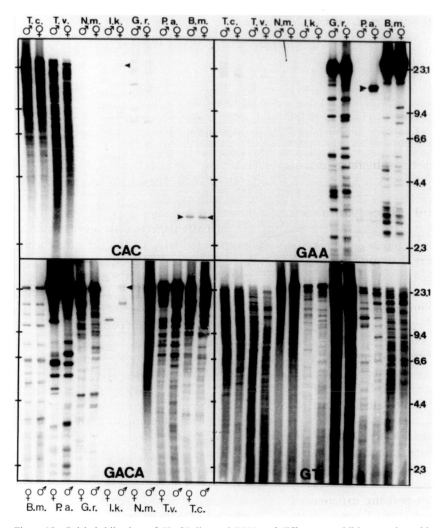

Figure 12. Gel hybridization of *Hinf* I-digested DNAs of different amphibian species with various simple repetitive sequences. (T.c.) *Triturus cristatus carnifex*, (T.v.) *Triturus vulgaris*, (N.m.) *Necturus maculosus*, (I.k.) *Ichthyophis kohtaoensis*, (G.r.) *Gastrotheca riobambae*, (P.a.) *Pyxicephalus adspersus*, and (B.m.) *Bufo melanostictus*. Note that apparent sex-specific hybridization can only be demonstrated in *Pyxicephalus adspersus* with the (GAA)₆ sequence.

It can be concluded that the Bkm sequences are not quantitatively associated with amphibian sex chromosomes as they are in snakes. A similar study in the fish species *Poecilia reticulata*, having almost homomorphic XY sex chromosomes, showed that the GATA/GACA simple repeats are quantitatively concentrated on the Y chromosome [91]. However, in other related poeciliid fishes these particular repeats are not sex-specific. Among many rodents, the GATA/GACA sequences are accumulated in the genome without displaying sex-specific hybridization signals [92]. In this context, the results obtained on the amphibian species are supporting the rule.

Sex-determining mechanisms in the Amphibia

Genetic or environmental sex determination?

The vast literature existing on sex-determination and sex-differentiation in the amphibians has already been subjected to exhaustive compilations. For those interested in an in-depth view of these studies, the reviews of Bull [61], Hayes [93] and Solari [94] are recommended.

In the extant vertebrates, there are at least two recognized mechanisms of sex determination: genetic sex determination (GSD) and environmental sex determination (ESD). In GSD one or more genes determine whether the primordial gonads develop into testes or ovaries, regardless of the external factors. In ESD each individual has the potential capacity to develop into a male or female, depending on environmental influences.

In some fish and lizards, many turtles, and all crocodilians the sex is determined by the incubation temperature of the eggs (ESD). Such a temperature-dependent sex determination was also suggested to operate in several amphibian species of the anuran genera *Bufo* and *Rana*, as well as in the urodelan genera *Pleurodeles* and *Hynobius* (for reviews, see [93, 95–98]). In all these studies it was shown that high or low temperatures of the rearing water alter the sex ratios of the developing larvae. Thus, in the studies performed on frogs, 100% males were obtained at high temperatures, whereas either 100% males or 100% females were produced in salamanders. However, the temperatures used (27–36°C) are normally not experienced by these species in the wild. Furthermore, when reared at a temperature experienced naturally by these species, a 1:1 sex ratio was obtained. Therefore, unlike in some fish and reptiles displaying ESD, temperature is not relevant for sex determination in amphibians. The role of temperature in sex determination of fish and reptiles is much more conceivable, because the effects already appear at temperatures within the ranges experienced by these species in their natural habitats. The experimental influences reported in amphibians most probably are artifacts of abnormally high temperatures at which the larvae were reared.

In a few individuals of some amphibian species, data have been collected which suggest the possibility of simultaneous or sequential hermaphroditism, as well as a spontaneous sex reversal in captivity [99, 100]. However, as argued by Hayes [93], there was either no histological examination of the gonads to support this conclusion, or the results may be explained by a meiotic exchange of sex-determining genes between the sex chromosomes in the progenitors of these individuals.

The record of XY or ZW sex chromosomes, regardless of their degree of morphological differentiation, is the decisive proof that a species possesess GSD. Although some 1000 species of the more than 4000 extant amphibians have been examined cytogenetically [1], only in about 5% of these species have morphologically distinguishable sex chromosomes been demonstrated [2–4]. However, with the application of the modern cytogenetic techniques it can expected that many further examples of differentiated sex chromosomes will be detected. Thus, the extensive analyses on the neotropical frog genus *Eleutherodactylus* carried out in our laboratory during the past ten years have already shown the presence of highly heteromorphic XY chromosomes in many populations of *E. johnstonei*, highly differentiated ZW sex chromosomes in all individuals of *E. shrevei* and *E. euphronides* examined, and Y-autosomal translocations (Robertsonian fusions) in a considerable number of the males in *E. riveroi* and *E. pulvinatus* [101]. In the karyotypes of frogs of the genus *Rana* occurring in Japan, many new fascinating forms of XY and ZW sex chromosomes have been discovered during recent years by means of thorough microscopical studies [35, 62–68, 102–104].

The apparent fact that the majority of amphibian species do not exhibit sex chromosomes visible at the microscopic level may indicate only that (1) more precise cytogenetic techniques have to be developed for their demonstration, or (2) that their sex chromosomes have not evolved to the extent where they are morphologically distinguishable. Although there may be many species with chromosomes having sex-determining genes, the differentiated regions are too small to be recognized cytologically.

Furthermore, as described above, many other experimental approaches that do not require morphological identification of the sex chromosomes have been successful in identifying the existence of GSD in the Amphibia. Thus, it can be concluded that under natural conditions, genetic sex determination (GSD) is operating in most (if not all) amphibian species. No definite proof for environmental sex determination (ESD) has been found in these vertebrates, but environmental factors may influence the genetic sex determination.

Sex-determining genes

ZFY sequences
The male-specific *ZFY* gene (Y-borne zinc finger protein) was originally cloned from the human Y chromosome [105, 106]. The *ZFY* probe detects male-specific sequences in the DNA of a range of eutherian mammals. A homologous sequence is located on the X chromosome of humans and mouse [105]. However, the autosomal location of *ZFY*-related sequences in the marsupials [107], as well as several observations in humans and mouse [108, 109] exclude *ZFY* as the testis-determining gene. Despite this

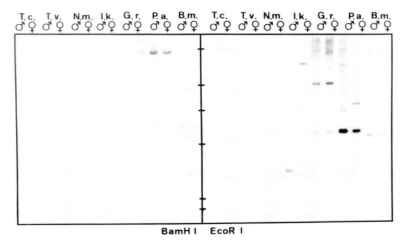

Figure 13. Autoradiograph of a Southern hybridization of *BamH* I- and *EcoR* I-digested genomic DNAs of (T.c.) *Triturus cristatus*, (T.v.) *Triturus vulgaris*, (N.m.) *Necturus maculosus*, (I.k.) *Ichthyophis kohtaoensis*, (G.r.) *Gastrotheca riobambae*, (P.a.) *Pyxicephalus adspersus*, and (B.m.) *Bufo melanostictus*, showing RFLPs for *ZFY* sequences. The filter was probed with *ZFY* (1.3 kb *Hind* III cross-hybridizing fragment), washed at 60°C in 0.3 M NaCl at room temperature, and exposed for five days with intensifying screen at −70°C.

controversy, *ZFY* may still be considered as an interesting sequence for the study of XY and ZW chromosomes in lower vertebrates. DNAs of several amphibian species with different sex chromosome constitutions (XX/XY, ZZ/ZW, and without identifiable sex chromosomes) were hybridized to the human *ZFY* sequence [4]. As shown in the Figure 13, *ZFY*-cross-hybridizing (homologous) sequences occur among the Anura in multiple copies. In *Gastrotheca riobambae* (XX/XY), no difference exists between the male and the female. On the contrary in *Pyxicephalus adspersus* (ZZ/ZW) and *Bufo melanostictus* (homomorphic ZW chromosomes) in addition to the homologous bands, certain bands differed between the sexes. This could be attributed perhaps to a restriction fragment length polymorphism (RFLP) for the *ZFY* locus. On the other side, the band at 7 kb in the female *Pyxicephalus adspersus* may well reside on the W chromosome. In *Leiopelma hochstetteri*, *Hynopius* sp., and *Dicamptodon tenebrosus*, identical hybridization patterns with *ZFY* have been found in both sexes [110]. A notable feature was found in *Ichthyophis kohtaoensis* (Apoda), where each sex possesses a single band that differ from each other in their molecular organization. As the sex chromosomal status is unknown in the Apoda, the *ZFY* probe may identify a sex-specific RFLP. These preliminary results in Amphibia stand in clear contrast to the data obtained in truly female (ZW) heterogametic vertebrates (birds and various reptile species) where *ZFY* reveals no male-female difference in the hybridization patterns [111].

SRY and SOX sequences

In mammals, sex determination is governed by a gene on the Y chromosome. In the presence of this gene, the embryonic gonad develops into a testis and the further development is male. In the absence of this gene, the embryonic gonad becomes an ovary and the further development is female. The responsible gene, referred to as TDF (testis-determining factor), is located in the short arm of the human Y chromosome, proximal to the pseudoautosomal boundary. This gene, called *SRY* (sex-determining region Y), was isolated by positional cloning from the critical sex-determining region on the Y chromosome, as defined by the minimal Y-specific segment present in XX males and absent in XY females [112]. *SRY* encodes a protein of 204 amino acid residues containing a conserved 79-amino acid DNA-binding and DNA-bending motif (HMG domain). The corresponding gene in the mouse, *Sry*, is located in the smallest region of the Y chromosome known to induce testicular differentiation, and is expressed in somatic cells of the gonadal ridge just before differentiation of the testis [113]. The ultimate proof that *SRY* is the TDF in mammals was provided by recording testis development in genetically female (XX) mice, experimentally made transgenic for the the *Sry* gene [114].

Although *SRY* is the primary male sex determinant in mammals, it is not clear whether it has a similar function, or even one that is sex-related, in the non-mammalian vertebrates. Thus, a probe from a conserved motif of the *SRY* gene was hybridized to male and female DNAs of 23 different species of fish, reptiles, and birds [115]. Hybridization occurred in species with male or female heterogamety, in species with and without sex chromosomes, and in those with temperature-dependent sex determination. No sex-specific signals were observed in this study. Furthermore, as yet no sex-specific hybridization of a *SRY* probe has been reported for any amphibian species.

SRY-related sequences are known as *SOX* (*SRY*-related HMG box) genes. A single study [116] reported identification of a *SOX* gene in the salamander *Pleurodeles waltl*, but its expression is not sex-specific in this species. In the frog *Rana rugosa*, *SOX9* was found to be expressed in both developing testes and ovaries, suggesting that this gene is involved in the gonadal development of male and female individuals [117]. This is dissimilar to the pattern found in mammals and birds, in which *SOX9* expression is male-specific. Finally, there are no indications that in *Xenopus laevis SOX* genes have a sex-specific expression pattern [118, 119].

Of all its relatives in the *SOX* gene family, *SRY* shows the most sequence similarity to *SOX3*. Actually, *SRY* genes of various species are more similar to *SOX3* than they are to each other, suggesting the possibility that *SRY* evolved from *SOX3* [121, 122]. The high conservation of *SOX3* between species indicates that it has a critical function in development. In mammals it is expressed in different tissues, including testis and the indifferent gonadal ridge [122, 123]. The homologous chicken c*SOX3* is restricted to the central nervous system. In *Xenopus laevis*, the homologue X*SOX3* is expressed exclusively in the ovary, and shows its highest expression early in oocyte development [124, 125].

Despite the evidence accumulated that GSD is operating in Anura and Urodela, there is little substantial information about how it functions. Although several DNA sequences homologous to the mammalian *ZFY*, *SRY* and *SOX* genes have been detected in the Anura or Urodela, none of these genes is an appropriate candidate to explain sex determination in these vertebrates.

Acknowledgements
Our studies on the Amphibia and all other vertebrate classes have been supported over many years by the Deutsche Forschungsgemeinschaft, the Commission of the European Communities, the Volkswagenstiftung, the Universitätsbund Würzburg, and the Deutscher Akademischer Austauschdienst. We are indebted to all colleagues who, during more than 20 years, supplied us with the many different species of vertebrates for cytogenetic and molecular analyses.

References

1 King M (1990) Amphibia. In: Animal cytogenetics, vol 4/2. pp 1–241, John B (ed), Gebrüder Borntraeger, Stuttgart
2 Schmid M (1983) Evolution of sex chromosomes and heterogametic systems in Amphibia. *Differentiation* 23 (Suppl): 13–22
3 Schmid M and Haaf T (1989) Origin and evolution of sex chromosomes in Amphibia: The cytogenetic data. In: Evolutionary mechanisms in sex determination, pp 37–56, Wachtel SS (ed), CRC Press, Boca Raton
4 Schmid M, Nanda I, Steinlein C, Kausch K, Haaf T and Epplen JT (1991) Sex-determining mechanisms and sex chromosomes in Amphibia. In: Amphibian cytogenetics and evolution, pp 393–430, Green DM, Sessions SK (eds), Academic Press, San Diego
5 Singh L (1974) Present status of sex chromosomes in amphibians. *Nucleus* 17: 17–27
6 Humphrey RR (1942) Sex of the offspring fathered by two *Amblystoma* females experimentally converted into males. *Anat Rec* 82 (Suppl 77): 469
7 Humphrey RR (1945) Sex determination in ambystomid salamanders: A study of the progeny of females experimentally converted into males. *Am J Anat* 76: 33–36
8 Humphrey RR (1957) Male homogamety in the Mexican axolotl: A study of the progeny obtained when germ cells of a genetic male are incorporated in a developing ovary. *J Exp Zool* 134: 91–101
9 Ohno S (1967) *Sex chromosomes and sex-linked genes*. Springer, Berlin
10 Chang CY and Witschi E (1955) Breeding of sex-reversed males of *Xenopus laevis* Daudin. *Proc Soc Exp Biol Med* 89: 150–152
11 Chang CY and Witschi E (1956) Gene control and hormonal reversal of sex differentiation in *Xenopus*. *Proc Soc Exp Biol Med* 93: 140–144
12 Gallien L (1953) Inversion totale du sexe chez *Xenopus laevis* Daud. à la suite d'un traitement gynogène par le benzoate d'oestradiol administré pendant la vie larvaire. *C R Acad Sci Ser D* 237: 1565–1566
13 Kawamura T and Nishioka M (1977) Aspects of the reproductive biology of Japanese anurans. In: The reproductive biology of amphibians, pp 103–139, Taylor DH, Guttman SI (eds), Plenum Press, New York
14 Wachtel SS and Ohno S (1979) The immunogenetics of sexual development. *Progr Med Genet* 3: 109–142
15 Wachtel SS, Koo GC, Boyse EA (1975) Evolutionary conservation of H-Y (male) antigen. *Nature* 254: 270–272
16 Wachtel SS, Wachtel GM, Nakamura D and Gilmour D (1983) H-Y antigen in the chicken. *Differentiation* 23 (Suppl): 107–115

17 Wolf U (1998) The serologically detected H-Y antigen revisited. *Cytogenet Cell Genet* 80: 232–235

18 Engel W and Schmid M (1981) H-Y antigen as a tool for the determination of the heterogametic sex in Amphibia. *Cytogenet Cell Genet* 30: 130–136

19 Zaborski P (1979) Sur la constance de l'expression de l'antigène H-Y chez le sexe hétérogamètique de quelques Amphibiens et sur la mise en évidence d'un dimorphisme sexuel de l'expression de cet antigène chez l'Amphibien Anoure *Pelodytes punctatus*. *C R Acad Sci Ser D* 289: 1153–1156

20 Hillis DM and Green DM (1990) Evolutionary changes of heterogametic sex in the phylogenetic history of amphibians. *J Evol Biol* 3: 49–64

21 Elinson RP (1983) Inheritance and expression of a sex-linked enzyme in the frog, *Rana clamitans*. *Biochem Genet* 21: 435–442

22 Schempp W and Schmid M (1981) Chromosome banding in Amphibia. VI. BrdU-replication patterns in Anura and demonstration of XX/XY sex chromosomes in *Rana esculenta*. *Chromosoma* 83: 697–710

23 Witschi E (1923) Ergebnisse der neueren Arbeiten über die Geschlechtschromosomen bei Amphibien. *Z Induk Abstamm Vererbungsl* 31: 287–312

24 Wright DA and Richards DM (1983) Two sex-linked loci in the leopard frog *Rana pipiens*. *Genetics* 103: 249–261

25 Witschi W (1929) Studies on sex differentiation and sex determination in amphibians: III. Rudimentary hermaphroditism and Y chromosome *Rana temporaria*. *J Exp Zool* 54: 157–223

26 Schmid M (1980) Chromosome banding in Amphibia. V. Highly differentiated ZW/ZZ sex chromosomes and exceptional genome size in *Pyxicephalus adspersus* (Anura, Ranidae). *Chromosoma* 80: 69–96

27 Schmid M and Bachmann K (1981) A frog with highly evolved sex chromosomes. *Experientia* 37: 242–244

28 Schmid M, Haaf T, Geile B and Sims S (1983) Chromosome banding in Amphibia. VIII. An unusual XY/XX-sex chromosome system in *Gastrotheca riobambae* (Anura, Hylidae). *Chromosoma* 88: 69–82

29 Schmid M, Haaf T, Geile B and Sims S (1983) Unusual heteromorphic sex chromosomes in a marsupial frog. *Experientia* 39: 1153–1155

30 Schmid M, Sims S, Haaf T and Macgregor HC (1986) Chromosome banding in Amphibia X. 18S and 28S ribosomal RNA genes, nucleolus organizers and nucleoli in *Gastrotheca riobambae*. *Chromosoma* 94: 139–145

31 Schmid M and Klett R (1994) Chromosome banding in Amphibia. XX. DNA replication patterns in *Gastrotheca riobambae* (Anura, Hylidae). *Cytogenet Cell Genet* 65: 122–126

32 Schmid M, Steinlein C, Friedl R, de Almeida CG, Haaf T, Hillis DM et al (1990) Chromosome banding in Amphibia. XV. Two types of Y chromosomes and heterochromatin hypervariability in *Gastrotheca pseustes* (Anura, Hylidae). *Chromosoma* 99: 413–423

33 Schmid M, Steinlein C, Feichtinger W, de Almeida CG, Duellman WE (1988) Chromosome banding in Amphibia. XIII. Sex chromosomes, heterochromatin and meiosis in marsupial frogs (Anura, Hylidae). *Chromosoma* 97: 33–42

34 Schmid M, Steinlein C and Feichtinger W (1989) Chromosome banding in Amphibia. XIV. The karyotype of *Centrolenella antisthenesi* (Anura, Centrolenidae). *Chromosoma* 97: 434–438

35 Schmid M, Ohta S, Steinlein C and Guttenbach M (1993) Chromosome banding in Amphibia. XIX. Primitive ZW/ZZ sex chromosomes in *Buergeria buergeri* (Anura, Rhacophoridae). *Cytogenet Cell Genet* 62: 238–246

36 Ponse K (1942) Sur la digamétie du crapaud hermaphrodite. *Rev Suisse Zool* 49: 185–189

37 Iturra P and Veloso A (1981) Evidence for heteromorphic sex chromosomes in male amphibians (Anura: Leptodactylidae). *Cytogenet Cell Genet* 31: 108–110

38 Schmid M, Steinlein C and Feichtinger W (1992) Chromosome banding in Amphibia. XVII. First demonstration of multiple sex chromosomes in amphibians: *Eleutherodactylus maussi* (Anura, Leptodactylidae). *Chromosoma* 101: 284–292

39 Mahony MJ (1991) Heteromorphic sex chromosomes in the Australian frog *Crinia bilingua* (Anura: Myobatrachidae). *Genome* 34: 334–337

40 Green DM (1988) Cytogenetics of the endemic New Zealand frog, *Leiopelma hochstetteri*: Extraordinary supernumerary chromosome variation and a unique sex-chromosome system. *Chromosoma* 97: 55–77

41 Green DM (1988) Heteromorphic sex chromosomes in the rare and primitive frog *Leiopelma hamiltoni* from New Zealand. *J Heredity* 79: 165–169

42 Sessions SK and Wiley JE (1985) Chromosome evolution in salamanders of the genus *Necturus*. *Brimleyana* 10: 37–52

43 Sessions SK (1980) Evidence for a highly differentiated sex chromosome heteromorphism in the salamander *Necturus maculosus* (Rafinesque). *Chromosoma* 77: 157–168

44 Lacroix J-C (1968) Étude descriptive des chromosomes en écouvillon dans le genre *Pleurodeles* (Amphibien, urodèle). *Ann Embryol Morphog* 1: 179–202

45 Lacroix J-C (1968) Variations expérimentales ou spontanées de la morphologie et de l'organisation des chromosomes en écouvillon dans le genre *Pleurodeles* (Amphibien, urodèle). *Ann Embryol Morphog* 1: 205–248

46 Lacroix J-C (1970) Mise en évidence sur les chromosomes en écouvillon de *Pleurodeles poireti* Gervais, amphibien urodèle, d'une structure liée au sexe, identifiant le bivalent sexual et marquant le chromosome W. *C R Acad Sci Ser D* 271: 102–104

47 Ferrier V, Jaylet A, Cayrol C, Gasser F and Buisan J-J (1980) Étude électrophorétique des peptidases érythrocytaires chez *Pleurodeles waltlii* (Amphibien Urodèle): Mise en evidence d'une liaison avec le sexe. *C R Acad Sci Ser D* 290: 571

48 Ferrier V, Gasser F, Jaylet A and Cayrol C (1983) A genetic study of various enzyme polymorphisms in *Pleurodeles waltlii* (urodele amphibian). II-peptidases: Demonstration of sex-linkage. *Biochem Genet* 21: 535–549

49 Schmid M, Olert J and Klett C (1979) Chromosome banding in Amphibia. III. Sex chromosomes in *Triturus*. *Chromosoma* 71: 29–55

50 Mancino G, Ragghianti M and Bucci-Innocenti S (1977) Cytotaxonomy and cytogenetics in European newt species. In: The reproductive biology of amphibians, pp 411–447, Taylor DH, Guttman SI (eds), Plenum Press, New York

51 Sims SH, Macgregor HC, Pellat PS and Horner HA (1984) Chromosome 1 in crested and marbled newts (*Triturus*): An extraordinary case of heteromorphism and independent chromosome evolution. *Chromosoma* 89: 169–185

52 Sessions SK and Kezer J (1987) Cytogenetic evolution in the plethodontid salamander genus *Aneides*. *Chromosoma* 95: 17–30

53 Kezer J and Macgregor HC (1971) A fresh look at meiosis and centromeric heterochromatin in the red-backed salamander, *Plethodon c. cinereus* (Green). *Chromosoma* 33: 146–166

54 León PE and Kezer J (1978) Localization of 5S RNA genes on chromosomes of plethodontid salamanders. *Chromosoma* 65: 213–230

55 Mancino G (1965) Osservazioni cariologiche sull' Urodelo della Sardegna *Euproctus platycephalus*: Morfologia dei bivalenti meiotici e dei lampbrush chromosomes. *Rend Accad Naz Lincei* 39: 540–548

56 Morescalchi A (1975) Chromosome evolution in the caudate Amphibia. In: Evolutionary biology, vol 8, pp 338–387, Dobzhansky T, Hecht MK, Steere WC (eds), Plenum Press, New York

57 Schmid M (1980) Chromosome evolution in Amphibia. In: Cytogenetics of vertebrates, pp 4–27, Müller H (ed), Birkhäuser, Basel

58 Nardi I, Andronico F, De Lucchini S and Batistoni R (1986) Cytogenetics of the European plethodontid salamanders of the genus *Hydromantes* (Amphibia, Urodela). *Chromosoma* 94: 377–388

59 Morescalchi A and Serra V (1974) DNA renaturation kinetics in some paedogenetic urodeles. *Experientia* 30: 487–489

60 Stephenson EM, Robinson ES and Stephenson NG (1972) Karyotype variation within the genus *Leiopelma* (Amphibia, Anura). *Can J Genet Cytol* 14: 691–702

61 Bull JJ (1983) *Evolution of sex-determining mechanisms*. Benjamin/Cumming, Menlo Park, California

62 Nishioka M, Miura I and Saitoh K (1993) Sex chromosomes of *Rana rugosa* with special reference to local differences in sex-determining mechanism. *Sci Rep Lab Amphibian Biol, Hiroshima Univ* 12: 55–81

63 Nishioka M, Kodama Y, Sumida M and Ryuzaki M (1993) Systematic evolution of 40 populations of *Rana rugosa* distributed in Japan elucidated by electrophoresis. *Sci Rep Lab Amphibian Biol, Hiroshima Univ* 12: 83–131

64 Nishioka M, Hanada H, Miura I and Ryuzaki M (1994) Four kinds of sex chromosomes in *Rana rugosa*. *Sci Rep Lab Amphibian Biol, Hiroshima Univ* 13: 1–34

65 Nishioka M and Hanada H (1994) Sex of reciprocal hybrids between the Murakami (ZZ-ZW type) population and Hamakita (XX-XY type) population in *Rana rugosa*. *Sci Rep Lab Amphibian Biol, Hiroshima Univ* 13: 35–50

66 Miura I, Ohtani H, Hanada H, Ichikawa Y, Kashiwagi A and Nakamura M (1997) Evidence for two successive pericentric inversions in sex lampbrush chromosomes of *Rana rugosa* (Anura: Ranidae). *Chromosoma* 106: 178–182

67 Miura I (1994) Sex chromosome differentiation in the Japanese brown frog, *Rana japonica*. I. Sex-related heteromorphism of the distribution pattern of constitutive hetero-chromatin in chromosome no. 4 of the Wakuya population. *Zool Sci* 11: 797–806

68 Miura I (1994) Sex chromosome differentiation in the Japanese brown frog, *Rana rugosa*. II. Sex-linkage analyses of the nucleolar organizer regions in chromosomes no. 4 of the Hiroshima and Saeki populations. *Zool Sci* 11: 807–815

69 Dournon C, Guillet F, Boucher D and Lacroix JC (1984) Cytogenetic and genetic evidence of male sexual inversion by heat treatment in the newt *Pleurodeles poireti*. *Chromosoma* 90: 261–264

70 Labrousse M, Guillemin C and Gallien L (1972) Mise en évidence sur les chromosomes de l'amphibien *Pleurodeles waltlii* Michah. de secteurs d'affinité différente pour le colo-rant de Giemsa à pH 9. *C R Acad Sci Ser D* 274: 1063–1065

71 Bailly S (1976) Localisation et signification des zones Q observées sur les chromosomes mitotiques de l'amphibien *Pleurodeles waltlii* Michah. après coloration par la moutarde de quinacrine. *Chromosoma* 54: 61–68

72 Beçak W, Beçak ML, Nazareth HRS and Ohno S (1964) Close karyological kinship between the reptilian suborder Serpentes and the class Aves. *Chromosoma* 15: 606–617

73 Singh L, Purdom IF and Jones KW (1976) Satellite DNAs and evolution of sex chromo-somes. *Chromosoma* 59: 43–62

74 Jones KW and Singh L (1981) Conserved repeated DNA sequences in vertebrate sex chromosomes. *Hum Genet* 58: 46–53

75 Jones KW (1983) Evolutionary conservation of sex-specific DNA sequences. *Differentia-tion* 23 (Suppl): 56–59

76 Jones KW (1984) The evolution of sex chromosomes and their consequences for the evo-lutionary process. In: Chromosomes today, vol 8, pp 241–255, Bennett MD, Gropp A, Wolf U (eds), Allen and Unwin, London

77 Green DM and Sharbel TF (1988) Comparative cytogenetics of the primitive frog, *Leio-pelma archeyi* (Anura, Leiopelmatidae). *Cytogenet Cell Genet* 47: 212–216

78 Baverstock PR, Adams M, Polkinghorne RW and Gelder M (1982) A sex-linked enzyme in birds: Z-chromosome conservation but no dosage compensation. *Nature* 296: 763–766

79 Schmid M, Enderle E, Schindler D and Schempp W (1989) Chromosome banding and DNA replication patterns in bird karyotypes. *Cytogenet Cell Genet* 52: 139–146

80 Nanda I, Shan Z, Schartl M, Burt DW, Koehler M, Nothwang H-G et al (1999) 300 million years of conserved synteny between chicken Z and human chromosome 9. *Nature Genet* 21: 258–259

81 Wright DA, Richards CM, Frost JS, Camozzi AM and Kunz BJ (1983) Genetic mapping in amphibians. In: Isozymes: Current topics in biological and medical research, vol 10, pp 287–311, Alan R Liss, New York

82 Elinson RP (1981) Genetic analysis of developmental arrest in an amphibian hybrid (*Rana catesbeiana, Rana clamitans*). *Dev Biol* 81: 167–176

83 Sumida M and Nishioka M (1994) Geographic variability of sex-linked loci in the Japanese brown frog *Rana japonica*. *Sci Rep Lab Amphibian Biol, Hiroshima Univ* 13: 173–195

84 Nishioka M and Sumida M (1994) The position of sex-determining genes in the chromo-somes of *Rana nigromaculata* and *Rana brevipoda*. *Sci Rep Lab Amphibian Biol, Hiro-shima Univ* 13: 51–97

85 Graf J-D (1989) Sex linkage of malic enzyme in *Xenopus laevis*. *Experientia* 45: 194–196

86 Graf J-D (1989) Genetic mapping in *Xenopus laevis*: Eight linkage groups established. *Genetics* 123: 389–398

87 Pardue ML and Gall J (1970) Chromosomal localization of mouse satellite DNA. *Science* 168: 1365–1368

88 Singh L, Purdom IF and Jones KW (1980) Chromosome satellite DNA. Evolution and conservation. *Chromosoma* 79: 137–157

89 Epplen JT, McCarry JR, Sutow S and Ohno S (1982) Base sequence of a cloned snake W chromosome fragment and identification of a male-putative mRNA in the mouse. *Proc Natl Acad Sci USA* 79: 3798–3802

90 Jones KW and Singh L (1985) Snakes and the evolution of sex chromosomes. *Trends Genet* 1: 55–61

91 Nanda I, Feichtinger W, Schmid M, Schröder JH, Zischler H and Epplen JT (1990) Simple repetitive sequences are associated with the differentiation of the sex chromosomes in the guppy fish. *J Mol Evol* 30: 456–462

92 Epplen JT (1988) On simple repeated GATA/GACA sequences in animal genomes: A critical reappraisal. *J Hered* 79: 409–417

93 Hayes TB (1998) Sex determination and primary sex differentiation in amphibians: Genetic and developmental mechanisms. *J Exp Zool* 281: 373–399

94 Solari AJ (1994) *Sex chromosomes and sex determination in vertebrates.* CRC Press, Boca Raton

95 Dodd JM (1960) Genetic and environmental aspects of sex determination in cold blooded vertebrates. *Mem Soc Endocrinol* 7: 17–44

96 Gallien L (1974) Intersexuality. In: Physiology of the Amphibia. Academic Press, New York, 523–549

97 Dournon C and Houillon C (1984) Démonstration génétique de l'inversion functionelle du phénotype sexuel femelle sous l'action de la température d'élevage chez l'amphibien urodéle: *Pleurodeles waltlii* Michah. *Reprod Nutr Dev* 24: 361–378

98 Dournon C, Houillon C and Pieau C (1990) Temperature sex-reversal in amphibians and reptiles. *Int J Dev Biol* 34: 81–92

99 Grafe TU, Linsenmair KE (1989) Protogynous sex change in the reed frog *Hyperolius viridiflavus. Copeia* 1989: 1024–1029

100 Collenot A, Durand D, Lauther M, Dorazi R, Lacroix J-C and Dournon C (1994) Spontaneous sex reversal in *Pleurodeles waltl* (urodele amphibia): Analysis of its inheritance. *Genet Res* 64: 43–50

101 Schmid et al; unpublished

102 Ohta S (1986) Sex determining mechanism in *Buergeria buergeri* (Schlegel). I. Heterozygosity of chromosome pair no. 7 in the female. *Sci rep Lab Amphibian Biol, Hiroshima Univ* 8: 29–43

103 Ohta S (1987) Sex determining mechanism in *Buergeria buergeri* (Schlegel). II. The effects of sex hormones on the differentiation of gonads and the offspring of sex-reversed females. *Sci Rep Lab Amphibian Biol, Hiroshima Univ* 9: 213–238

104 Ohta S, Sumida M and Nishioka M (1999) Sex-determining mechanism in *Buergeria buergeri* (Anura, Rhacophoridae). III. Does the ZZW triploid frog become female or male? *J Exp Zool* 283: 295–306

105 Page DC, Mosher R, Simpson EM, Fisher EMC, Mardon G, Pollack J et al (1987) The sex-determining region of the human Y-chromosome encodes a finger protein. *Cell* 51: 1091–1164

106 Page DC (1988) Is ZFY the sex-determining gene on the human Y-chromosome? *Philos Trans R Soc London* 322: 155–157

107 Sinclair AH, Foster JW, Spencer JA, Page DC, Palmer M, Goodfellow PN et al (1988) Sequences homologous to ZFY, a candidate human sex-determining gene, are autosomal in marsupials. *Nature* 336: 780–783

108 Palmer MS, Sinclair AH, Berta P, Ellis NA, Goodfellow PN, Abbas NE et al (1989) Genetic evidence that ZFY is not the testis-determining factor. *Nature* 342: 937–939

109 Koopman P, Gubbay J, Collignon J and Lovell-Badge R (1989) Zfy gene expression patterns are not compatible with a primary role in mouse sex determination. *Nature* 342: 940–942

110 Zeyl CW, Green DM and Nishioka Y; unpublished

111 Bull JJ, Hillis DM and O'Steen S (1988) Mammalian ZFY sequences exist in reptiles regardless of sex-determining mechanisms. *Science* 242: 567–569

112 Sinclair AH, Berta P, Palmer MS, Hawkins JR, Griffiths Bl, Smith MJ et al (1990) A gene from the human sex-determining region encodes a protein with homology to a conserved DNA-binding motif. *Nature* 346: 240–244

113 Koopman P, Münsterberg A, Capel B, Vivian N and Lovell-Badge R (1990) Expression of a candidate sex-determining gene during mouse testis differentiation. *Nature* 348: 450–452

114 Koopman P, Gubbay J, Vivian N, Goodfellow PN and Lovell-Badge R (1991) Male development of chromosomally female mice transgenic for *Sry. Nature* 351: 117–121

115 Tiersch TR, Mitchell MJ and Wachtel SS (1991) Studies on the phylogenetic conservation of the *SRY* gene. *Hum Genet* 87: 571–573

116 Chardard D, Chesnel A, Gozc C, Dournon C and Berta P (1993) PW *Sox*-1: The first member of the *SOX* gene family in urodeles. *Nucleic Acids Res* 21: 3576–3578

117 Takase M, Noguchi S and Nakamura M (2000) Two *Sox*9 messenger RNA isoforms: Isolation of cDNAs and their expresion during gonadal development in the frog *Rana rugosa. FEBS Lett* 466: 249–254

118 Miyata S, Miyashita K and Hosoyama Y (1996) *SRY*-related genes in *Xenopus* oocytes. *Biochim Biophys Acta* 1308: 23–27

119 Hiraoka Y, Komatsu N, Sakai Y, Ogawa M, Shiozawa M and Aiso S (1997) XLS 13A and XLS13B: *SRY*-related genes of *Xenopus laevis. Gene* 197: 65–71

120 Schmid M and Steinlein C (1991) Chromosome banding in Amphibia. XVI. High-resolution replication banding patterns in *Xenopus laevis. Chromosoma* 101: 123–132

121 Foster JW and Graves JAM (1994) An *SRY*-related sequence on the marsupial X chromosome: Implications for the evolution of the mammalian testis-determining gene. *Proc Natl Acad Sci USA* 91: 1927–1931

122 Stenovic M, Lovell-Badge R, Collignon J and Goodfellow PN (1993) *SOX3* is an X-linked gene related to *SRY. Hum Mol Genet* 2: 2013–2018

123 Collignon J, Sockanathan S, Hacker A, Cohen-Tannoudji M, Norris D and Rastan S (1996) A comparison of the properties of *SOX3* with SRY and two related genes, *SOX1* and *SOX2. Development* 122: 509–520

124 Koyano S, Ito M, Takamatsu N, Takiguchi S and Shiba T (1997) The *Xenopus SOX3* gene expressed in oocytes of early stages. *Gene* 188: 101–107

125 Penzel R, Oschwald R, Chen YL, Tacke L and Grunz H (1997) Characterisation and early embryonic expression of a neural specific transcription factor *XSOX3* in *Xenopus laevis. Int J Dev Biol* 41: 667–677

Genes and Mechanisms in Vertebrate Sex Determination
ed. by G. Scherer and M. Schmid
© 2001 Birkhäuser Verlag Basel/Switzerland

Endocrine and environmental aspects of sex differentiation in gonochoristic fish

Jean-François Baroiller[1],* and Yann Guiguen[2],*

[1] *CIRAD-EMVT (Centre International en Recherche Agronomique pour le Développement) and* [2] *INRA (Institut National de la Recherche Agronomique), Laboratoire de Physiologie des Poissons, Campus de Beaulieu, 35042 Rennes Cedex, France*

Summary. This paper reviews current knowledge concerning the endocrine and environmental regulation of gonadal sex differentiation in gonochoristic fish. In gonochoristic fish, although potentially active around this period, the hypothalamo-pituitary axis is probably not involved in triggering sex differentiation. Although steroids and steroidogenic enzymes are probably not the initial triggers of sex differentiation, new data, including molecular approaches, have confirmed that they are key physiological steps in the regulation of this process. Environmental factors can strongly influence sex differentiation in gonochoristic fish. The most important environmental determinant of sex would appear to be temperature. Interactions between environmental factors and genotype have been suggested for gonochoristic fish.

Introduction

With over 20000 species, the class of fishes exhibits a large variety of adaptation responses to match the vast array of existing ecological habitats. The present review will focus on the teleost super-order which constitutes more than 90% of the fish species.

One of the most intriguing phenomena is probably the large number of reproductive strategies developed by these species. Sexuality represents an important aspect of these reproductive strategies: "members belonging to the class Pisces exemplify an almost complete range of various types of sexuality from synchronous hermaphroditism, protandrous and proto-gynous hermaproditism, to gonochorism" [1].

In addition to the diversity of sexuality in fishes, it is now well establish-ed that phenotypic sex in fish may depend on external factors, although the effect of such factors will differ from one species to the next [2, 3]. This plasticity of gonadal development in fish, which contrasts with the more stable patterns found in higher vertebrates, has given rise to a number of exciting questions concerning both its adaptive significance [4, 5] and the underlying genetic and physiological regulations involved.

At this point, we must refer to Hayes' terminology which places emphasis on the difference between "sex determination" and "sex differentiation":

---

* Both J.F. Baroiller and Y. Guiguen contributed equally to this work.

the former expression designates mechanisms that direct sex differentiation, while the latter refers to the development of testes or ovaries from the undifferentiated or bipotential gonad [6]. In the present review, genetic sex determination (GSD) will be only briefly evoked.

In gonochoristic species presenting a simple heterogametic GSD model (XX/XY or ZW/ZZ systems) the possibility of hormonal sex inversion has been used to obtain genetic monosex male or female populations. Not only XX neomales or ZZ neofemales, but also new viable genotypes (YY males and WW females, by crossing XY neofemales with XY males or ZW neomales with ZW females, respectively) can be used as breeders for producing monosex populations [7, 8]. These possibilities have relevant practical implications for fish farming [9, 10] and the hormonal induction of sex inversion has now been extended to a large number of species. In addition, genetic monosex populations are tremendously helpful in biological studies as they allow research to be carried out on a large number of known-sex individuals for which the sex is known from fertilisation, and thus before any detectable sign of sexual differentiation. Examples of such studies will be given in the following review concerning both the effects of external factors on sex determination and the physiological investigations of sex differentiation.

Several comprehensive reviews dealing with the morphological and histological description of sex differentiation [11] and sex inversion [2] have already been published. These aspects will therefore not be developed in the present paper in order that we may focus on two active fields of research for which a wealth of recent data have been obtained, namely the role of external factors and the involvement of endocrine factors in gonochoristic sex differentiation. It should be noted, however, that the chronology of gonadal sex differentiation is highly variable from one species to the next, but also within a given species where the growth rate (and thus water temperature) is an important factor [12]. In light of this, the time scale adopted for the kinetics of gonadal development is often expressed in terms of the age of larvae in days multiplied by degrees post-fertilization (PF) instead of only in days PF. The actual kinetics of gonadal differentiation should thus be determined prior to each new experiment where an accurate correlation between morphological features and physiological events is to be analyzed. Unfortunately, there is still a lack of definitive criteria for the detection of the very first discrete signs of differentiation [12]. Until now, the initiation of meiotic activity, shortly after proliferation of active germ cells, has frequently been reported as being the first recognizable indication of ovarian differentiation. This is due to the precocious differentiation of female germ cells which contrasts with the late appearance of the first meiotic prophase feature in future spermatocytes.

Genetic sex determination

Although some species show morphologically well differentiated sex chromosomes [13], cytogenetic examinations are rarely helpful in identifying sex chromosomes in fish due to the low occurrence of heteromorphy [14], even if some species show morphologically well differentiated sex chromosomes. Various methodologies, however, have been conclusive in the identification of GSD which can hardly be considered primitive [7, 15, 16]. The possibility of obtaining viable gynogenetic, androgenetic or hormonally inversed individuals offers original and powerful tools for such studies [17].

No simple model of GSD can be generalised in fish. Both male heterogamety (XX/XY), such as in several salmonids, and female heterogamety (ZZ/ZW), in *Gambusia*, have been reported. On the other hand, male and female heterogamety may be found even in the same species, as shown in the platyfish, *Xiphophorus*, which could be an argument for a recent divergence of sex chromosomes [15, 18]. Other species-specific models have been proposed, incorporating multiple sex chromosomes, polygenic sex determination and autosomal influence [8, 15, 19]. Conversely, knowledge concerning the genetic determination of hermaphroditism remains limited. Molecular studies focusing on genes involved in fish GSD have been recently developed. *SRY*-type HMG (High Mobility Group) box (*Sox*) genes have been identified in some fish species [20, 21]. However, like other lower vertebrates, no functional sex determining gene equivalent to *SRY* has been demonstrated. Only sex-specific probes have been obtained, all of them being very specific to each species and even sometimes to a particular population of fish [13, 16, 22, 23].

Endocrine regulation of sex differentiation

Involvement of the central nervous system (CNS)

In gonochoristic species, very few studies have addressed the role of the CNS in the process of gonadal sex differentiation. In addition, these studies have focused on the involvement of the hypothalamo-pituitary system *via* gonadotropin releasing hormone (GnRH) and gonadotropin[1] hormones (GTHs). Using specific antibodies, GTH I immunoreactive cells have been found in rainbow trout, *Oncorhynchus mykiss*, pituitary when mitosis of germ cells are first detected in the gonads [24], and localized in the Pars

---

[1] We will rely in this review on the classical terminology for fish GTHs (GTH I and GTH II) although a new nomenclature has been proposed recently in fish. This new terminology proposes that GTH I (Gonadotropin type I) be referred to as FSH (Follicular Stimulating Hormone) and GTH II to as LH (Luteinizing Hormone). Unless stated otherwise in the article, we use the generic term GTH.

Proximal Distalis (PPD) close to hatching (but no difference was detected in the pattern of expression of these cells in rainbow trout between male and female) [25]. In a closely related salmonid, the Coho salmon, *O. kisutch*, GTH I immunoreactive cells were detected two weeks after hatching [26]. When specifically searched for, GTH II was not detected during the time surrounding the sex differentiation period either in rainbow trout [24, 25] or Coho salmon [26]. Using non-specific antibodies (either non-specific according to the species or the type of GTH), GTH immunoreactive cells were detected in the PPD of rainbow trout, when the gonads were thought to be still undifferentiated histologically [27]. In the common carp, *Cyprinus carpio*, GTH-containing cells were found in the pituitary gland of three week-old larvae [28]. GnRH immunoreactivity was only investigated in rainbow trout around the time of sex differentiation and was detected in both sexes in several brain regions including those where GTH I was localized [25]. Growth hormone (GH) immunoreactivity in the pituitary gland of rainbow trout was detected at an even earlier stage than GTH I, namely, slightly before the appearance of Primordial germ cells (PGC) [24]. It was suggested that GH may be involved in the sex differentiation process as it is known that GH can act on regulation of steroidogenesis in adult salmonid gonads. In carp, a precocious sex differentiation was achieved following treatment of four week old larvae with homologous pituitary extracts [29]. This treatment resulted in a considerable increase in the PGC number, gonad size and GTH plasma levels as well as causing, in some fish, an acceleration in the appearance of histological features characteristic of female differentiation [29]. In light of the major role of GTHs in regulating gonadal steroidogenesis in adults [30], Fitzpatrick and collaborators [31] have examined the potential effects of partially purified salmon GTH, on gonads or gonadal-kidney explant steroidogenesis during sex differentiation in rainbow trout. GTH is shown to stimulate gonadal steroid production but only after histological gonadal differentiation. A much earlier stimulation of androstenedione[2] (Δ_4) production by the anterior kidney (interrenal) is also observed. Whether this production is significant with respect to gonadal differentiation in fish is not known although a participation of the interrenal in the production of steroids potentially acting on gonad differentiation has been proposed in rainbow trout [32]. All these studies have demonstrated that the hypothalamo-pituitary axis is potentially active around the time of sex differentiation and that steroids can have feedback effects on this axis in much the same way as it does in adults [33]. Although a pituitary synthesis of GTH I seems to be clearly established around the time of gonadal differentiation, at least in salmonids, it remains to be determined whether there is an active secretion.

---

[2] Common names, abbreviations used, and systematic names for the steroids cited in the text are given in Table 1.

Table 1. Common/systematic names of the main steroids with the abbreviations used

Common names	Abbreviations	Systematic name
Estrogens (C₁₈)		
Estradiol-17β	E_2	1,3,5,(10)-estratriene-3,17β-diol
Estrone	E_1	3-hydroxy-1,3,5,(10)-estratriene-17-one
Ethynyl-estradiol	EE_2	1,3,5(10)-estratrien-17α-ethynyl-3,17β-diol
Androgens (C₁₉)		
Androstenedione	Δ_4	4-androstene-3,17-dione
Testosterone	T	17β-hydroxy-4-androsten-3-one
Dehydroepiandrosterone	DHEA	3β-hydroxy-5-androsten-17-one
5α-Dihydrotestosterone	DHT	17β-hydroxy-5α-androstan-3-one
11-oxygenated androgens		
11β-Hydroxyandrostenedione	11βOHΔ₄	11β-hydroxy-4-androstene-3,17-dione
11β-Hydroxytestosterone	11βOHT	11β,17β-dihydroxy-4-androsten-3-one
11-ketotestosterone	11KT	17β-hydroxy-4-androstene-11,17-dione
Adrenosterone	Ad	4-androstene-3,11,17-trione
Synthetic androgens		
17α-Methyltestosterone	MT	4-androstene-17α-methyl-17β-ol-3-one
17α-Methyldihydrotestosterone	MDHT	5α-androstan-17α-methyl-17β-ol-3-one
17α-Ethynyltestosterone		4-androstene-17α-ethynyl-17β-ol-3-one
Progestagens (C₂₁)		
Progesterone	P_4	4-pregnene-3,20-dione
Pregnenolone	P_5	3β-hydroxy-5-pregnen-20-one
17-Hydroxyprogesterone	17P_4	17-hydroxy-4-pregnene-3,20-dione
17-Hydroxypregnenolone	17P_5	3β,17-dihydroxy-5-pregnen-20-one
17,20β-Dihydroxyprogesterone	DHP	17,20β-dihydroxy-4-pregnen-3-one
Corticosteroids (C₂₁)		
Cortisol		11β,17,21-trihydroxy-4-pregnene-3,20-dione
Cortisone		17,21-Dihydroxy-4-pregnene-3,11,20-trione
Cholesterol (C₂₇)		5-cholesten-3β-ol

Involvement of steroid hormones

Parabiose experiments

In the medaka, trunks of newly hatched fry were transplanted into the anterior eye chamber of adult animals of both sexes. When transferred into a male host, the gonads of a genetic female graft develop into an abnormal gonad structure containing spermatogenetic cells. When transferred into a female host, the gonads of a genetic male graft still develop into testis [34]. It was concluded that differentiation of male germ cells requires "male sex hormones" while differentiation of female germ cells is not induced by physiological levels of "female sex hormones". Among these putative "sex hormones", steroids have received the most attention.

Involvement of steroid hormones in gonochoristic species

Present knowledge concerning the role of steroid hormones in the process of sex differentiation has been mainly acquired through indirect techniques such as the use of treatments with steroid hormones, steroid enzyme inhibitors or steroid receptor antagonists. The data obtained from these experiments are only exploitable when an effect on the resulting sex-ratio is found, as an absence of effect can be due to an inadequate timing, dosage or mode of administration of the compounds. Moreover, most of the inhibitors or antagonist molecules used in fish have only been proven to be active in mammals. Measuring steroid hormone levels during gonadal differentiation or describing the steroid enzyme potentialities in the differentiated gonads also provides important information. In addition, steroid producing cells have been identified during the process of sex differentiation in a few species [review in 11].

Steroid treatments

Yamamoto postulated in 1969 that steroids were the natural "sex inducers", estrogens being the "gynoinducers" and androgens being the "androinducers" [1]. From that moment on, a vast number of experiments were conducted dealing with steroid treatments in fish. The aim of most of these treatments has been to control the sex phenotype in fish species of commercial interest [8]. Thus, only studies which provide some physiological insight as to the regulation of sex differentiation by steroid hormones will be examined here.

First of all, it should be specified that the gonadal sex phenotype in gonochoristic fish can only be manipulated around the time of the sex differentiation period. However, in the adult goldfish *Carassius auratus*, 11-ketotestosterone (11KT) implant treatments in incompletely ovariectomized females can lead to the development of testicular tissue, suggesting that germ cells can still retain a certain level of bi-potentiality [35].

Apart from the numerous treatments that are known classically to be effective in inducing feminization (using estrogens) or masculinization (using androgens), there exist a number of paradoxical results on the feminizing effects of androgens in fish [36, 37]. In an all-female genotypic population of Chinook salmon *O. tshawytscha*, balneation treatments were performed with either the synthetic, aromatisable methyltestosterone (MT) or the synthetic, non-aromatisable androgens methyldehydrotestosterone (MDHT), and the relative masculinizing potencies were MDHT > MT. Thus, it seems that the aromatisation of androgens decreases their relative masculinizing potencies [38]. However, in the channel catfish, *Ictalurus punctatus*, and the blue catfish, *Ictalurus furcatus*, paradoxical feminizations have also been observed following treatments with non-aromatisable androgens [39] and these types of puzzling feminization effects remain to be explained.

In the Coho salmon, two well differentiated periods of steroid sensitivity have been demonstrated with a maximum efficiency of feminization with estradiol-17β (E$_2$) balneation treatment carried out one day after hatching and a maximum efficiency of masculinization with MT one week later [40]. Although the effects of steroid treatments on masculinization or feminization have been well documented, very little data exists concerning the mechanism of action of these steroids. For instance, aromatisation is thought to decrease the masculinization efficiency of some aromatisable androgens [37, 38], but whether this is due to a decrease in androgen concentrations or an increase in estrogen concentrations is unknown. The physiological effects of steroid treatments in the differentiation of gonads has received very little attention. Selective incorporation and accumulation of radio-labelled steroids have been demonstrated in the gonads of medaka [41]. In rainbow trout, E$_2$ or MT treatments performed on all-male and all-female populations at the onset of the first feeding inhibited the *in vitro* production of steroids in both sexes, and did not induce the gonadal steroid secretion pattern of the opposite sex [31]. In the common carp, treatments with E$_2$ failed to feminize XY males, but produced intersexed animals on XY$_{(mas/mas)}$ males (animals homozygous for a recessive mutation in a gene called *mas* for mas-culinization). Together with the fact that these XX$_{(mas/mas)}$ males produced estrogens (but not normal XY males), it seems that increasing endogenous estrogen levels drive differentiation slightly towards female differentiation but without totally overriding the masculinizing effects of endogenous 11-oxygenated androgens [42].

Apart from these common treatments, a few studies have shown some effects of corticosteroids [43] or progestins [44] on sex differentiation. For corticosteroids this effect could be explained by a conversion into 11-oxygenated androgens [43].

In vivo treatments with steroid enzyme inhibitors
In the salmon *O. tschawytscha* balneation treatment of all-female populations with a non-steroidal aromatase inhibitor (AI) resulted in induced masculinization and increased masculinization induced by a low dosage of MT. Thus AI, by reducing aromatisation of MT, greatly enhances the masculinizing effect of an aromatisable androgen [45]. In rainbow trout and in tilapia, *O. niloticus*, treatment with the steroidal aromatase inhibitor 1,4,6-androstatriene-3-17-dione (ATD) resulted in masculinization of all-female populations [46–49]. High dosage treatments of rainbow trout with cyanoketone, a 3β-hydroxysteroid dehydrogenase (3βHSD) inhibitor, resulted in a significant increase in males. Surprisingly, this treatment also resulted in an important increase in the 3βHSD activity in the interrenal tissue, but not in the gonads of treated fish. This could result from an indirect stimulation of steroidogenesis in the interrenal tissues producing 11βOHΔ_4 able to masculinize the gonads [32]. However, treatment with metopyrone, a 11β-hydroxylase (11βH) inhibitor, failed to induce feminization, although the inhibition of 11βH was found to be ineffective, at least in the gonads [48, 50].

In vivo treatments with steroid receptor antagonists
Nearly all attempts to masculinize or feminize fish using steroid receptor
antagonists have failed [46–48, 51]. Only one unpublished report [cited in
51] mentions that the androgen receptor antagonist, cyproterone acetate,
induces feminization in the medaka. For estrogen receptor antagonism, one
experiment reports a masculinizing effect of tamoxifen in a hybrid population of tilapia, *O. niloticus* X *O. aureus* [52].

In vitro steroid metabolism
Some steroidogenic potentialities of eggs or very young embryos have
been shown in different species of salmonids [53, 54]. These pathways are
thought to be mainly for the deactivations of active steroids of maternal
origin [54]. Few studies have investigated the steroidogenic potentialities
of the differentiating gonads during the period encompassing sex differentiation. In the tilapia *O. niloticus*, aromatase activity was specific to ovaries
shortly after differentiation whereas $11\beta H$ and 11β-hydroxysteroid dehydrogenase ($11\beta HSD$) activities were specific to testis [50]. The same
pattern has been observed during gonadal differentiation in the rainbow
trout ([48] and author's unpublished results; see Fig. 1). In the same species,
the undifferentiated gonads have been shown to possess $3\beta HSD$ and 17-
hydroxylase (17H) activities. Shortly after gonadal differentiation, testes
are also able to synthesize 11β-hydroxyandrostenedione ($11\beta OH\Delta_4$) and
adrenosterone (Ad). In females, there is no potential for 11-oxygenated
androgen synthesis and estrogen synthesis potentiality is only acquired
much later following gonadal differentiation [55, 56]. In larvae of the catfish, *C. gariepinus*, 11-oxygenated androgens were also specific for testis
differentiation [57]. In the common carp, the major metabolites detected in
XY males are 11-oxygenated androgens, with nearly no estrogens, whereas
in $XY_{(max/+)}$ females, estrogens are the major metabolites with no 11-oxy-
genated androgens produced at all. In $XX_{(mas/mas)}$ males, both estrogens
and 11-oxygenated androgens are produced and animals differentiate into
the male phenotype. In conclusion, it would appear that the precocious
synthesis of 11-oxygenated androgens in the $XX_{(mas/mas)}$ animals directs
male testicular differentiation [42].

Steroid assays
Because of the very small size of differentiating gonads in fish, many
studies carried out to measure steroids during sex differentiation have been
performed on eggs, embryos, or wholebody extracts. Most of these studies
have come to the conclusion that steroids, some of them probably of maternal origin, can be detected and metabolized very early [53, 54, 58]. Interpretation of these experiments is difficult since extra-gonadal steroid production sites exist in fish [59], and differences between sexes have rarely
been found [60]. However, using all-male and all-female populations
of rainbow throut it has been demonstrated that higher levels of Δ_4 and

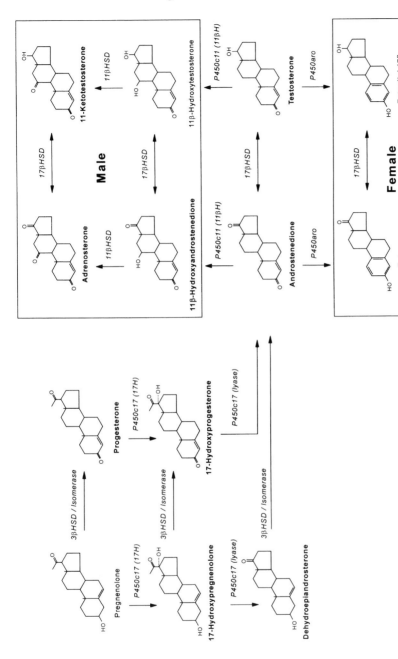

Figure 1. Schematic representation of some gonadal steroidogenetic pathways in fish and identified steroids (in bold type) following *in vitro* steroid metabolism with pregnenolone and androstenedione precursors in rainbow trout and/or tilapia differentiating gonads (authors' unpublished results). Enzyme activities are P450aro = aromatase, P450c11 (11βH) = 11β-Hydroxylase, P450c17 (17H) = 17-Hydroxylase, P450c17 (lyase) = 17,20-lyase, HSD = hydroxysteroid dehydrogenase.

testosterone (T) are produced in the testis, two weeks before histological gonadal differentiation. Gonadal secretion of E_2 is higher in female gonads but only after histological gonadal differentiation [31]. In *O. niloticus*, T levels are only detectable in the gonads and serum after testicular differentiation [61]. Also, a transient peak of T and 11KT has been detected in gonads at 22 dpf and a peak of E_2 during ovarian differentiation [62].

Steroid enzyme immunodetection and gene expression
In the eel *Anguilla anguilla*, cholesterol side chain cleavage cytochrome ($P450_{scc}$) was immunocytochemically detected in gonads before sex differentiation using a rat antibody [63]. During the process of sex differentiation in *O. niloticus*, two patterns of immunoreactivity were detected in gonads, one with a positive staining for all antibodies tested ($P450_{scc}$, $3\beta HSD$, cytochrome 17-hydroxylase/17,20 lyase = $P450_{c17}$, aromatase = $P450_{aro}$) and one with no staining at all. These positive staining reactions became detectable a few days before morphological sex differentiation and increased thereafter. Based on the positive staining for $P450_{aro}$, these gonads were assumed to be female. Weak positive reactions were detected at 30 dpf in presumed testis, but $P450_{aro}$ immunostaining was never detected [11]. Also, in the rainbow trout, expression of the *$P450_{aro}$* gene is specifically detected and highly expressed only in female gonads two weeks before the first sign of histological differentiation [46–48] (Fig. 2). Conversely, the *$P450_{c11}$* gene (11β-hydroxylase) is highly over-expressed in male gonads as early as the specific $P450_{aro}$ expression found in female gonads [48] (Fig. 2).

Steroid receptivity
Although steroid enzyme potentialities have been studied in some fish, the problem concerning the receptivity of the differentiating gonads has only occasionally been addressed. In rainbow trout, estrogen and androgen receptor genes were expressed early in both male and female gonads before and during sex differentiation, showing that gonadal receptivity for gonadal steroids may be acquired by both sexes [46–48] (Fig. 2).

Involvement of others factors
Genes previously found to be important in the process of gonadal sex differentiation in mammals have also been found in fish such as the Wilms' Tumor predisposition gene (*WT1*), which has been cloned in the pufferfish *Fugu rubripes* (Miles et al., unpublished sequence, Accession number AL021531) and in the zebrafish *Brachydanio rerio* [64]. The Steroidogenic Factor 1 (*SF1*), or *SF1* homologues belonging to the *FTZ-F1* family, have also been found both in the zebrafish [65] and rainbow trout [66]. Finally, a *Sox9* gene [20] and some *SRY*-related sequences [21] have been described in rainbow trout. None of these genes have been studied with respect to gonadal differentiation events, however.

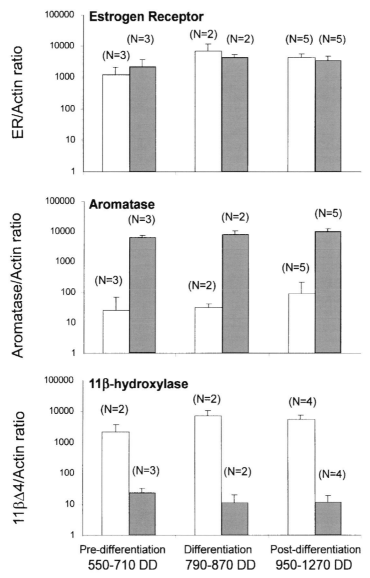

Figure 2. Semi-quantification by RT-PCR (Reverse-Transcription Polymerase-Chain Reaction) of some genes expression in gonads of rainbow trout before (pre-differentiation), during (differentiation, i.e., first oocytes meiosis in females), and after (post-differentiation) the histological gonadal differentiation (authors' unpublished results). Males are represented by open bars and females by black solid bars. Data are represented as ratio of the gene of interest over β-actin gene and expressed with a logarithmic scale. DD = Degree × Days. Details on the semi-quantitative technique have been published in [47].

Finally, two sequence homologues to *zona pellucida* proteins ZP2 and ZP3 were shown to be differentially expressed in the medaka gonads starting at five days post-hatching [67].

Influence of environmental factors on sex differentiation

Environmental factors influencing gonadal sex differentiation

In gonochoristic species, sex differentiation can be influenced by exogenous factors. Most studies have focused on the effec of exogenous steroids on sex differentiation. Although the influence of social factors in the regulation of sex inversions in certain hermaphroditic species has been well described, the presentation of indisputable evidence concerning the effect of environmental factors on sex differentiation in gonochoristic fish is more recent. As a result, very few environmental factors have been studied in only a limited number of species. Moreover, it is important to note that a number of studies concerned with the effect of environmental factors on sex differentiation have used, deliberately or not, inbred lines of fish species. This is particularly true for those works using species from the poecilids family as laboratory models. The relative importance of such inbreeding on the sensitivity to environmental factors remains to be determined [7]. In that regard, inbreeding has been shown to induce spontaneous XX male production (fixation of rare masculinizing recessive alleles) at least in the common carp [68] and the rainbow trout (E. Quillet, unpublished data). In gonochoristic fish species, as in other vertebrates displaying environmental sex determination (ESD), the main environmental determinant of sex may be the prevailing temperature during early determination [69]. This also happens to be the factor which has received the greatest attention. Most of the studies on temperature effects have focused on reptiles and amphibians: temperature effects, also called temperature sex determination or TSD have been demonstrated in a wide variety of crocodiles, turtles, lizards and certain species of frogs and salamanders [6, 69]. In amphibians, these temperature effects may only be artefacts due to the abnormally high rearing temperatures used and thus, under natural conditions, sex could depend exclusively on genetic factors [6]. In fish, the putative effects of other factors such as pH, salinity, photoperiod, or social interactions have received much less attention.

Effect of pH

The strong influence of pH, either alone or through an interaction with temperature, has been reported in a number of species. In *Xiphophorus helleri*, the development of male monosex populations (100%) or nearly female monosex populations (< 2% males) is obtained at an acid pH (6.2) or at a slightly basic value (7.8), respectively [70]. Similar results have been described in another poecilid, *Poecilia melanogaster*, and in 7 out of 37 species of *Apistogramma* (Cichlids) under study: male proportions are inversely proportional to pH [70, 71]. In *A. caetei*, for example, balanced populations are observed at an acid pH (53–60% males at pH 4.5–5.5), whereas under more neutral conditions, almost exclusively female populations are produced (4% males at pH 6.5).

Density
In the paradise fish *Macropodus opercularis* [72], individual isolation favours testicular differentiation (89% males), whereas grouping individuals induces a female differentiation which is proportional to the density (25% at the lowest density, 66% females at the highest density).

Relative size of juveniles
In the Midas cichlid, *Cichlasoma citrinellum*, stable size ranks are reported within a cohort from the juvenile stage to sexual maturity. Within a cohort of siblings, the larger individuals differentiate into males whereas smaller members undergo female differentiation. Males are always larger than females and have the primary role of defending the breeding territory. Finally, females prefer large males. The experimental alteration of size distribution (grading according to size) within a group results in a slight deviation in favour of females in both initially lower and upper size groups; assuming that sex is exclusively determined by genetic factors, both male (fish above the median size) and female (below the median size) groups should be obtained. Therefore, sex differentiation can be influenced by relative size [73].

Effect of temperature on sex differentiation: different models of thermosensitivity

In gonochoristic fish, although the thermosensitivity of gonadal sex differentiation is being discovered in an increasing number of species (as more research is carried out on this topic), it is probably premature to attempt to define TSD patterns. In an effort to summarise the main results, however, three main response types have been retained here:

1. In most of the species identified as thermosensitive, such as *Hoplosternum littorale* [74], *P. lucida* [75] and *P. melanogaster* [71], the proportion of males increases with temperature and/or female differentiation is favoured by low temperatures (Fig. 3). In the tilapias *O. niloticus* [76–79], *O. aureus* [80] and the red tilapia (a 4-way-hybrid) from the Red Florida strain [81], low temperatures do not affect the sex ratio. Conversely, in the Atherinids *Menidia menidia* [82–84], *Patagonina hatcheri* [85–86], *Odonthestes bonariensis* [86, 87], and *O. argentinensis* [85], predominantly female populations are produced at low temperatures. Finally, it can be stated that, contrary to TSD patterns in reptile

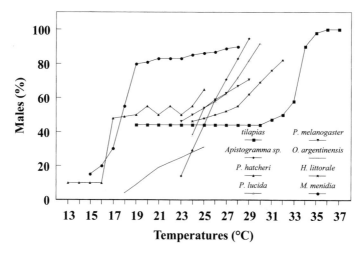

Figure 3. Masculinisation by high temperatures and/or feminisation by low temperatures in several fish species with temperature-dependent sex determination. Schematic representation of the extreme data from [71, 74–87].

species [69], a complete change from monosex female to monosex male populations at extreme temperatures is never observed, with the exception of *Odonthestes bonariensis* [86, 87]. This last species differs from reptiles in that it lacks the abrupt threshold of TSD observed in the latter vertebrates [86].

2. In only two species it has been suggested that high temperatures may favour ovarian differentiation and/or low temperatures induce testicular differentiation (Fig. 4). Indeed, monosex male populations of the sea bass *Dicentrarchus labrax* can be obtained at low temperatures (15°C), whereas variable but low proportions of females (< 27% females) are produced under more typical rearing conditions (24–25°C) [88]. A slight feminization (63% females) is induced by high temperatures (34°C) in the catfish *Ictalurus punctatus* [89], whereas low and intermediate temperatures do not affect the typical balanced sex ratio (1 : 1). It should be noted, however, that these last two studies were carried out on a limited number of progenies.

3. Finally, only one species the hirame *Paralichthys olivaceus* is currently known to produce monosex male populations at both high and low temperatures (U-shape curve) [90]. At intermediate temperatures, the observed sex proportions fit well with a model of female homogamety (XX/XY): genetically mixed populations (from pairs of XX females and XY males) generate balanced sex ratios whereas monosex female populations are sired by XX[3] males (Fig. 5).

---

[3] These males were produced through functional masculinization, and identified by their monosex female progenies: XX × XX → 100% female XX.

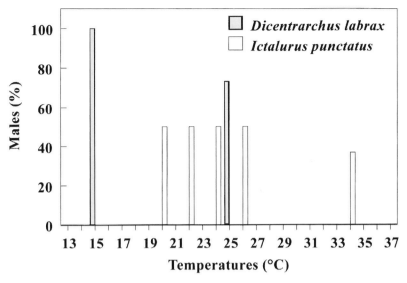

Figure 4. Feminisation by high temperatures and/or masculinisation by low temperatures in the European sea bass, *Dicentrarchus labrax*, and the catfish, *Ictalurus punctatus*. Schematic representation of data from [88–89].

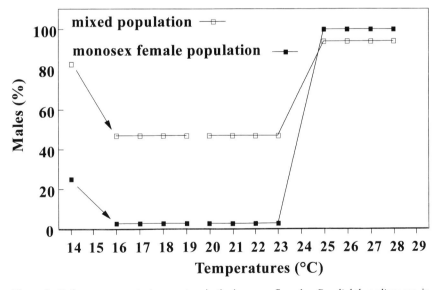

Figure 5. U shape response to temperature in the japanese flounder, *Paralichthys olivaceus*, in a mixed population and a genetically all-female population. Schematic representation of data from [90].

Thermosensitive period

In the tilapias [76, 77], some Atherinids [86, 87, 91] and the sea bass [88], thermal treatments must begin before the onset of histological gonadal sex differentiation and must also at least partially overlap this critical period. In the above mentioned species, as is true in reptiles, this period is very similar to the hormono-sensitive period, during which a hormonal treatment must be applied if such a treatment is to be efficient (Fig. 2).

Genotype/temperature interactions

Species identified as thermosensitive can present all the possible genetic sex determination systems already described in fish [15]. For example, in tilapias[4] [92], American catfish[5] [89] and hirame[5] [90] simple male or female homogamety[6] has been described whereas more complex sex determinations (sex chromosomes cannot be identified even through indirect approaches) based on several major sex determining factors are reported in the Atlantic silverside, *M. menidia* [82].

  In most thermosensitive fish species, interactions between environmental factors and genotype have been strongly suggested:

1. In *M. menidia*, a highly variable thermosensitivity has been reported both between and within populations. A sex-linked growth dimorphism has been described in favour of females. Females stemming from the very first low temperature reproduction will benefit from a longer growth season. As size preferentially favours the relative fecundity of females as compared to males, populations at different latitudes will compensate the thermal and seasonal differences by adjusting the magnitude of their response to temperature, and therefore adjust their relative thermosensitivity. As southern populations benefit from a long reproductive season (and subsequently a long growing season for the future generation), they present a high degree of thermosensitivity and mostly female populations are produced. Conversely, northern populations are faced with a short reproductive season, and are not, or only weakly, thermosensitive [82, 84, 93]. Moreover, a high variability in thermosensitivity is also described within a given family [82].
2. In the tilapia *O. niloticus*, within the same strain, the most and least sensitive genetic female progenies (sired by XX males), will respectively produce 98% and 5% males at high masculinizing temperatures (36°C) vs. 100% and 96% females at control temperatures (27–28°C is generally considered to be the optimal temperature for this species) [77]. Very similar

---

[4] XX/XY in *O. niloticus*, and ZZ/ZW in *O. aureus*.
[5] *Ictalurus punctatus* and *P. olivaceus* both exhibit a female homogamety XX/XY.
[6] A pair of homologous sexual chromosomes in the male (ZZ/ZW) or the female (XX/XY).

results are observed using progenies sired from genetic XY males (Fig. 3), and male proportions increase by 18–44% following masculinizing high-temperature treatments [76, 78]. Finally, parental effects are strongly suggested in certain tilapia species [81; Baroiller and Clota, unpublished data]. A given female can successively mate with two different males, and respectively produce highly sensitive (100% and 75% males at 36°C and 27°C respectively) and non-sensitive (40% and 41% males at 36°C and 27°C respectively) progenies. Similar results have been obtained with the progenies from two different females successively mated with the same male (61% and 82% males at 36°C). Conversely, successive progenies from a same breeder pair always present similar thermosensitivities (58–65% males at 27°C/80–83% at 36°C).

3. Conversely, in another thermosensitive tilapia species, *O. aureus*, temperature/genotype interactions seem to be weak and systematically all-male or almost monosex male progenies (95–100%) are produced following high temperature treatment [80]. In *P. lucida*, only one out of two homozygous lines is thermosensitive. Segregation studies between the two lines suggest dominance of the thermosensitive trait [75]. In another poecilid, the guppy *P. reticulata*, an XX line seems to be more sensitive to environmental factors than a classic line due to fluctuating season-dependent sex ratios [94].

For species in which the effect of other environmental factors has been proven, similar genotype interactions have also been described. In the paradise fish, *M. opercularis*, sex differentiation can be influenced by density as has already been reported above [72]. After genetic selection of dominance in five successive generations, the positive line (where the dominance trait has been positively selected) presents a 77% proportion of males vs. 52% in the control line.

Absence of temperature-dependent sex determination in fish

In light of the adaptive significance of TSD demonstrated in *M. menidia*, the hypothesis that other fish with similar life histories could also present TSD has been tested in two cyprinodontid fish, *Cyprinodon variegatus* and *Fundulus heteroclitus* [95]. Contrary to the expected hypothesis, low temperatures do not produce female offspring in *F. heteroclitus* nor do they generate male progenies in *C. variegatus* (43 to 58% females at 15, 18, 21 and 28°C). Although the existence of TSD has not been demonstrated in the studied populations, the authors cautioned against drawing definitive conclusions for both species. Indeed, intraspecific variations in sex determining mechanisms have been reported in several fish species [18–19]. In the rainbow trout *Oncorhynchus mykiss*, the influence of temperature on sex differentiation has been investigated during the early developmental stages. High temperature treatments (22 to 29°C) applied soon after hatching and of various duration (10 min to 14 days) do not affect the sex ratio [96]. More recently, longer high temperature (19°C) treatments have been applied on genetic all-male and all-female populations during their entire hormonal sensitive period, and this without any effect on the respective sex ratios (Baroiller and Guiguen, unpublished data).

Genetic evidence of a functional sex inversion by temperature in gonochoristic fish

Temperature-induced sex masculinization has been demonstrated in sensitive tilapia progenies (79% males at 36°C/46% males at 27°C): after reaching sexual maturity, high temperature treated (HTT) and control males were individually progeny tested. All-female (or almost monosex female) progenies were only detected in the HTT group, whereas progenies with balanced sex ratios were produced in both groups. The former all-female progenies were sired by XX males (genetic females functionally sex inverted to males by high temperatures); conversely, balanced progenies resulted from mating using a genetic XY male [76, 78].

Involvement of estrogens in the mechanisms of thermosensitivity

In tilapia progenies submitted to masculinizing temperature treatments, lower levels of *aromatase* gene expression are reported [97]. Therefore, at least in tilapia, temperature appears to modulate the expression of key steroidogenic enzymes.

Discussion/Conclusion

Information on the role of the CNS on gonadal sex differentiation in lower vertebrates is scarce. In amphibians, GTHs do not seem to be involved in this process [6] and in mammals it is generally thought that sexualisation of the CNS is a secondary event controlled by factors secreted by the already differentiated gonads [98]. In gonochoristic fish, the CNS does not seem to have a preponderant role, at least as an initial trigger, but the hypothalamo-pituitary axis is potentially active at about the time differentiation takes place and thus may be needed for the completion of the sex differentiation process. In reptiles, however, it should be noted that a hypothesis has been proposed implicating the brain expression of aromatase enzyme in the process of gonadal sex differentiation [99]. In many lower vertebrates, steroids seem to play a crucial role in the process of gonadal sex differentiation [1, 69, 100]. More specifically, aromatase is now considered to be a key enzyme in gonadal sex differentiation at least in reptiles [69, 100] and birds [101]. Two types of theoretical models have been proposed based either on the presence/absence of aromatase enzyme or on the androgen to estrogen ratio and thus 5α-reductase to aromatase activity [99, 100]. In fish, the importance of estrogens in gonadal differentiation can also be considered and the enzyme aromatase is probably one of the key enzymes needed for ovarian differentiation in gonochoristic fish. Despite this preponderant role of aromatase, however, results concerning 11-oxygenated androgens in fish also deserve attention and we would like to propose a parallel hypothesis to the one suggested in reptiles [99, 100]. 11-oxygenated androgens are the active androgens in fish [102–103]. Thus, instead of the androgen to estrogen ratio adopted for reptiles, the 11-oxygenated androgen to estrogen ratio in fish would direct either male (excess 11-oxygenated androgens) or female (excess estrogens) differentiation. This hypothesis is in agreement with most of the information on steroids and steroid enzyme activity and/or gene expression described to date concerning fish gonadal sex differentiation.

Concerning the other factors involved in gonadal sex differentiation, a few studies have shown that the genes implicated in mammalian differentiation are also found in lower vertebrates. In that regard, the *Sox9* and *WT1* genes have been cloned and studied in reptiles [104] and birds [64, 105, 106], and the expression profiles of these genes during gonadal differentiation are consistent with those obtained in mammals. The antimüllerian hormone (AMH) cDNA has also been cloned in the chicken [107] and turtle [108], and the pattern of expression in chick gonads is strongly dimorphic during sex differentiation as previously found in mammals [106]. Some of these genes thought to be important in the process of mammalian gonadal sex differentiation have also been found in fish, but none of them have been studied with respect to gonadal differentiation events. With the advent of molecular tools, this area will probably be investigated very soon.

Since the first evidence of thermosensitivity in *M. menidia* [83], various studies have clearly demonstrated that environmental factors can strongly influence sex differentiation and thus, the sex ratios in gonochoristic fish. In most of the sensitive fish species examined, interactions between environmental factors and genotype have been suggested [72, 82, 76–81, 84, 93, 94, 109]. Such environmental effects could at least partially explain some of the unexpected sex ratios recorded in fish [110]. As is seen in reptiles [69, 99], three main patterns of thermosensitivity seem to exist in sensitive fish. Contrary to reptile patterns, however, and with the exception of *O. bonariensis* [86], all-male and/or all-female populations are generally not produced at extreme temperatures; this could result from strong temperature/genotype interactions. In certain thermosensitive species, breeding pairs may thus generate highly sensitive progenies with 100% male progenies in the treated group, as has been seen for tilapia (Baroiller and Clota, unpublished data). The characteristics of thermosensitivity also

196   J.-F. Baroiller and Y. Guiguen

differ substantially from those observed in reptiles. Thus, in fish, there is a
large range of temperatures over which the populations sex ratios fit well
with a strict GSD model (this is especially true of certain thermosensitive
species, such as the hirame *P. olivaceus*, which follow a male or male homogamety model [90]). Conversely, in reptiles, both sexes are only obtained
within a narrow transitional range (less than 1°C in some species) [69]. In
this respect, thermosensitivity in fish resembles that in amphibians such as
Pleurodeles waltl and *P. poireti*. In these pleurodeles species, genetic sex
determination (male homogamety ZZ/ZW) governs sex differentiation at
ambient temperatures (balanced sex ratios are observed at 16 to 24°C),
whereas warm and cold temperatures strongly affect gonadal sex differentiation [111]. In addition, genotype/temperature interactions are rarely
reported in amphibians [111].

Several characteristics of fish sexuality are worth being reminded: (1) the
ability to manipulate the phenotypic sex differentiation in gonochoristic or
the onset of sex inversion in hermaphroditic species by hormonal or environmental treatments, (2) the ability to obtain new viable and fertile
sexual genotypes in gonochoristic species (YY, WW), and thus produce
genetically all-male and all-female populations, and (3) the ability to
observe both homogametic and heterogametic systems within the same
gender (tilapia) or even the same species (the platyfish). All of these features make the fish a tremendous model for the study of sex determination
and gonadal sex differentiation in vertebrates.

1 Yamamoto T (1969) Sex differentiation. In: Fish Physiology, vol III, pp 117–175, Hoar WS and Randall EJ (eds), Academic Press, New York
2 Chan STH and Yeung WSB (1983) Sex control and sex reversal in fish under natural con-ditions. In: Fish Physiology, vol IXB, pp 171–222, Hoar WS, Randall DJ and Donaldson EM (eds), Academic Press, New York
3 Francis RC (1992) Sexual lability in teleost: developmental factors. *Quart Rev Biol* 67: 1–17
4 Conover DO (1984) Adaptative significance of temperature-dependent sex determination in a fish. *Am Nat* 123: 298–313
5 Warner RR (1988) Sex change in fishes: hypotheses, evidences and objections. *Env Biol Fish.* 22: 81–90
6 Hayes TB (1998) Sex determination and primary sex differentiation in amphibians: genetic and developmental mechanisms. *J Exp Zool* 281: 373–399
7 Chevassus B, Devaux A, Chourrout D and Jalabert B (1988) Production of YY rainbow trout males by self-fertilization of induced hermaphrodites. *J Hered* 79: 89–92
8 Hunter GA and Donaldson EM (1983) Hormonal sex control and its application to fish culture. In: Fish Physiology, vol IXb, pp 223–291, Hoar WS and Randall DJ (eds), Academic Press, New York
9 Chevassus B, Chourrout D and Jalabert B (1979) Le contrôle de la reproduction chez les poissons. I. Les populations "monosexes". *Bull Fr Piscic* 274: 18–31
10 Guiguen Y, Baroiller JF, Jalabert B and Fostier A (1996) Le contrôle du sexe phénotypique chez les poissons. *Pisc Fr* 124: 16–19
11 Nakamura M, Kobayashi T, Chang X and Nagahama Y (1998) Gonadal sex differentiation in teleost fish. *J Exp Zool* 281: 362–372

12 Bruslé J and Bruslé S (1982) La gonadogenèse des poissons. *Reprod Nutr Dévelop* 22: 453–491

13 Reed KM, Bohlander SK and Phillips RB (1995) Microdissection of the Y chromosome and fluorescence *in situ* hybridization analysis of the sex chromosomes of lake trout, *Salvelinus namaycush. Chromosome Res* 3: 221–226

14 Beçak W (1983) Evolution and differentiation of sex chromosomes in lower vertebrates. *Differentiation* 23 (Suppl): S3–S12

15 Chourrout D (1988) Revue sur le déterminisme génétique du sexe des poissons téléostéens. *Bull Soc Zool Fr* 113: 123–144

16 Nakayama I, Foresti F, Tewari R, Schartl M and Chourout D (1994) Sex chromosome polymorphism and heterogametic males revealed by two cloned DNA probes in the ZW/ZZ fish, *Leporinus elongatus. Chromosoma* 103: 31–39

17 Thorgaard GH (1983) Chromosome set manipulation and sex control in fish. In: Fish Physiology, vol IXB, pp 405–434, Hoar WS and Randall DJ (eds), Academic Press, New York

18 Kallman KD (1984) A new look at sex determination in poeciliid fishes. In: Evolutionary genetics of fishes, pp 95–171, Turner BJ (ed), Plenum Publishing Corporation, New York

19 Price DJ (1984) Genetics of sex determination in fishes: a brief review. In: Fish Reproduction: strategies and tactics, pp 77–89, Pottsand GW and Wootton RJ (eds), Academic Press, London

20 Takamatsu N, Kanda H, Ito M, Yamashita A, Yamashita S and Shiba T (1997) Rainbow trout SOX9: cDNA cloning, gene structure and expression. *Gene* 202: 167–170

21 Fukada S, Tanaka M, Iwaya M, Nakajima M and Nagahama Y (1995) The Sox gene family and its expression during embryogenesis in the teleost fish, medaka (*Oryzias latipes*). *Develop Growth Differ* 37: 379–385

22 Nanda I, Schartl M, Feichtinger W, Epplen J and Schmid M (1992) Early stages of sex chromosome differentiation in fish analysed by single repetitive DNA sequences. *Chromosoma* 101: 301–310

23 Du SJ, Devlin RH and Hew CL (1993) Genomic structure of growth hormone genes in chinook salmon (*Oncorhynchus tshawytscha*): presence of two functional genes, GH-I and GH-II, and a male-specific pseudogene, GH-Ψ. *DNA and Cell Biol* 12(8): 739–751

24 Saga T, Oota Y, Nozaki M and Swanson P (1993) Salmonid pituitary gonadotrophs. III. Chronological appearance of GTH I and other adenohypophysioal hormones in the pituitary of the developing rainbow trout (*Oncorhynchus mykiss irideus*). *Gen Comp Endocrinol* 92: 233–241

25 Feist G and Schreck CB (1996) Brain-pituitary-gonadalaxis during early development and sexual differentiation in the rainbow trout, *Oncorhynchus mykiss. Gen Comp Endocrinol* 102: 394–409

26 Mal AQ, Swanson P and Dickhoff WW (1989) Immunocytochemistry of the developing salmon pituitary gland. *Am Zool* 29: 94A

27 Van Den Hurk R (1982) Effects of steroids on gonadotropic (GTH) cells in the pituitary of rainbow trout, *Salmo gairdneri*, shortly after hatching. *Cell Tissue Res* 237: 285–289

28 Van Winkoop A, Timmermans LPM and Booms GHR (1987) The expression of germ cell differentiation antigens, as defined with monoclonal antibodies, in correlation with the ontogeny of gonadotropic cells in the hypophysis of carp. In: Proceeding of the Third International Symposium on Reproductive Physiology of Fish, pp 222, Idler DR, Crim LW and Walsh JM (eds) St John's Newfoundland, August 2–7, 1987

29 Van Winkoop A, Timmermans LPM and Goos HJTh (1994) Stimulation of gonadal and germ cell development in larval and juvenile carp (*Cyprinus carpio* L.) by homologous pituitary extract. *Fish Physiol Biochem* 13(2): 161–171

30 Suzuki K, Nagahama Y and Kawauchi H (1988) Steroidogenic activities of two distinct salmon gonadotropins. *Gen Comp Endocrinol* 71: 452–458

31 Fitzpatrick MS, Pereira CB and Schreck CB (1993) *In vitro* steroid secretion during early development of mono-sex rainbow trout: sex differences, onset of pituitary control, and effects of dietary steroid treatment. *Gen Comp Endocrinol* 91: 199–215

32 Van den Hurk R and Leeman WR (1984) Increase of steroid-producing cells in interrenal tissue and masculinization of gonads after long-term treatment of juvenile rainbow trout with cyanoketone. *Cell Tissue Res* 237: 285–289

33 Kah O, Anglade I, Leprétre E, Dubourg P and de Monbrison D (1993) The reproductive brain in fish. *Fish Physiol Biochem* 11: 85–98

34 Satoh N (1973) Sex differentiation of the gonad of fry transplanted into the anterior chamber of the adult eye in the teleost, *Oryzias latipes. J Embryol Exp Morph* 30(2): 345–358

35 Kobayashi M, Aida K and Stacey NE (1991) Induction of testis development by implantation of 11-ketotestosterone in female goldfish. *Zool Sci* 8: 389–393

36 Goudie CA, Redner BD, Simco BA and Davis KB (1983) Feminization of channel cat fish by oral administration of steroid sex hormones. *Trans Amer Fish Soc* 112: 670–672

37 Piferrer F and Donaldson EM (1991) Dosage-dependent differences in the effect of aromatizable and non-aromatizable androgens on the resulting phenotype of coho salmon (*Oncorhynchus kisutch*). *Fish Physiol Biochem* 9(2): 145–150

38 Piferrer F, Baker IJ and Donaldson EM (1993) Effects of natural, synthetic, aromatizable, and non-aromatizable androgen in inducing male sex differentiation in genotypic female chinook salmon (*Oncorhynchus tshawytscha*). *Gen Comp Endocrinol* 91: 59–65

39 Davis KB, Goudie CA, Simco BA, Tiersch TC and Carmichael GJ (1992) Influence of dihydrotestosterone on sex determination in channel catfish and blue catfish: period of developmental sensitivity. *Gen Comp Endocrinol* 86: 147–151

40 Piferrer F and Donaldson EM (1987) Influence of estrogen, aromatizable and nonaromatizable androgen during ontogenesis on sex differentiation in coho salmon (*Oncorhynchus kisutch*). In: Proceeding of the Third International Symposium on Reproductive Physiology of Fish, pp 135, Idler DR, Crim LW and Walsh JM (eds), St John's, Newfoundland, August 2–7, 1987

41 Hishida TO (1965) Accumulation of estrone-16-C^{14} and diethylstilbestrol-(monoethyl-1-C^{14}) in larval gonads of the medaka, *Oryzias latipes*, and determination of the minimum dosage of estrogen for sex reversal. *Gen Comp Endocrinol* 5: 137–144

42 Komen J, Lambert JGD, Richter CJJ and Goos HJTh (1995) Endocrine control of sex differentiation in XX female, and in XY and XX male common carp (*Cyprinus carpio*, L). In: Proceeding of the Fifth International Symposium on the Reproductive Physiology of Fish, pp 383, Goetz F and Thomas P (eds), Austin, Texas, 2–8 July 1995

43 Van den Hurk R and Van Oordt PGWJ (1985) Effects of natural androgens and corticosteroids on gonad differentiation in the rainbow trout, *Salmo gairdneri. Gen Comp Endocrinol* 57: 216–222

44 Van Den Hurk R and Sloft GA (1981) A morphological and experimental study of gonadal sex differentiation in rainbow trout, *Salmo gairdneri. Cell Tissue Res* 218: 487–497

45 Pifferer F, Zanuy S, Carrillo M, Solar II, Delvin RH and Donaldson EM (1994) Brief treatment with aromatase inhibitor during sex differentiation causes chromosomally female salmon to develop as normal, function males. *J Exp Zool* 270: 255–262

46 Guiguen, Y, Ricordel MJ and Fostier A (1998) Involvement of estrogens in the process of sex differentiation in rainbow trout, (*Oncorhynchus mykiss*): *in vivo* treatments, aromatase activity, and aromatase gene expression. *J Exp Zool* 281: 506 (abstract)

47 Guiguen, Y, Baroiller JF, Ricordel MJ, Iseki K, McMeel OM, Martin SAM et al (1999) Involvement of estrogens in the process of sex differentiation in two fish species: the rainbow trout (*Oncorhynchus mykiss*) and a tilapia (*Oreochromis niloticus*). *Mol Reprod Dev* 54: 154–162

48 Guiguen, Y, Govoroun M, D'Cotta H, McMeel OM and Fostier A (1999) Steroids and gonadal sex differentiation in the rainbow trout, *Oncorhynvhus mykiss*. In: Proceeding of the 6th International Symposium on Reproductive Physiology of Fish, Bergen, Norway, 5–9 July 1999

49 Maléjac ML (1993) Etude des stéroides sexuels comme facteurs physiologiques de la différenciation sexuelle chez la truite arc-en-ciel (*Oncorhynchus mykiss*). Phd dissertation. University of Rennes I. p 129

50 Baroiller JF, Guiguen, Y, Iseki K and Fostier A (1998) Physiological role of androgens on gonadal sex differentiation in two teleost fish, *Oncorhynchus mykiss* and *Oreochromis niloticus. J Exp Zool* 281: 506–507 (abstract)

51 Schreck CB (1974) Hormonal treatment and sex manipulation in fishes. In: Control of sex in fishes, pp 84–106, Schreck CB (ed), Extension division, Virginia polytechnical institute and state university, Blacksburg

52 Hines GA and Watts SA (1995) Non-steroidal chemical sex manipulation of tilapia. *J World Aquac Soc* 26(1): 98–102

53 Khan MN, Renaud RL and Leatherland JF (1997) Steroid metabolism by embryonic tissues of arctic charr, *Salvelinus alpinus. Gen Comp Endocrinol* 105: 344–357

54 Yeoh CG, Schreck CB and Feist GW (1996) Endogenous steroid metabolism is indicated by fluctuations of endogenous steroid and steroid glucuronide levels in early development of the steelhead trout (*Oncorhynchus mykiss*). *Gen Comp Endocrinol* 103: 107–114

55 Van den Hurk R, Lambert JGD and Peute J (1982) Steroidogenesis in the gonads of rainbow trout fry (*Salmo gairdneri*) before and after the onset of gonadal sex differentiation. *Reprod Nutr Dévelop* 22(2): 413–425

56 Van den Hurk R, Slof GA and Schurer FA (1980) Gonadal sex differentiation in rainbow trout, *Salmo gairdneri*, with special reference to the effects of steroid hormones and N,N-Dimethylformamide. *Gen Comp Endocrinol* 40: 323

57 Van den Hurk R, Richter CJJ and Janssen-Dommerholt J (1989) Effects of 17α-methyltestosterone and 11β-hydroxyandrostenedione on gonad differentiation in the african catfish, *Clarias gariepinus. Aquaculture* 83: 179–191

58 Schreck CB, Fitzpatrick MS, Feist GW and Yeoh CG (1991) Steroids: developmental continuum between mother and offspring. In: Proceedings of the Fourth International Symposium on the Reproductive Physiology of Fish, pp 256–258, Scott AP, Sumpter JP, Kime DE and Rolfe MS (eds), Norwich, UK 7–12 July 1991

59 Watts SA, Wasson KM and Hines GA (1995) A sexual paradox: androgen and estrogen synthesis in the tilapia kidney. In: Proceeding of the Fifth International Symposium on the Reproductive Physiology of Fish, p 277, Goetz F and Thomas P (eds), Austin, Texas, 2–8 July 1995

60 Feist G, Schreck CB, Fitzpatrick MS and Redding JR (1990) Sex steroid profiles of coho salmon (*Oncorhynchus kisutch*) during early development and sexual differentiation. *Gen Comp Endocrinol* 80: 299–313

61 Nakamura M ad Nagahama Y (1989) Differentiation and development of leydig cells, and changes of testosterone levels during testicular differentiation in tilapia *Oreochromis niloticus. Fish Physiol Biochem* 7(1–4): 211–219

62 Hines GA, Wibbels T and Watts SA (1998) Sex steroid levels and steroid metabolism in relation to early gonadal development in normal and sex-reversed tilapia. *J Exp Zool* 281: 521 (abstract)

63 Eckstein B, Kedar H, Castel M and Cohen S (1988) The appearance of cytochrome P-450 cholesterol side chain cleavage enzyme in the differentiating gonad of the European eel (*Anguilla anguilla* L.). In: Reproduction in Fish – Basic and applied aspects in endocrinology and genetics, no 44, pp 147–152, INRA (ed), les colloques de l'INRA

64 Kent J, Coriat AM, Sharpe PT, Hastie ND and van Heyningen V (1995) The evolution of WT1 sequence and expression pattern in the vertebrates. *Oncogene* 11: 1781–1792

65 Liu D, Le Drean Y, Ekker M, Xiong F and Hew L (1997) Teleost FTZ-F1 homolog and its splicing variant determine the expression of the salmon gonadotropin II beta subunit gene. *Mol Endocrinol* 11: 877–890

66 Ito M, Masuda A, Yumoto K, Otomo A, Takahashi N, Takamatsu N et al (1998) cDNA cloning of a new member of the FTZF1 subfamily from a rainbow trout. *Biochim Biophys Acta* 1395: 271–274

67 Kanamori A (1997) Gene expression during sex differentiation in medaka, *Oryzias latipes*. In: Abstracts of the XIII International Congress of Comparative Endocrinology, pp 585–590. Yokohama, Japan. 16–21 November 1997

68 Komen J, De Boer P and Richter CJJ (1992) Male sex reversal in gynogenetic XX females of common carp (*Cyprinus carpio* L.) by a recessive mutation in a sex determining gene. *J Heredity* 83: 431–434

69 Pieau C (1996) Temperature variation and sex determination inreptiles. *Bioassays* 18(1): 19–26

70 Rubin DA (1985) Effect of pH on sex ratio in cichlids and a poeciliid (Teleostei). *Copeia* 233–235

71 Römer U and Beisenherz W (1996) Environmental determination of sex in *Apistogramma* (Cichlidae) and two other freshwater fishes (Teleostei). *J Fish Biol* 48: 714–725

72 Francis RC (1984) The effects of bidirectional selection for social dominance on agonistic behaviour and sex ratios in the paradise fish, (*Macropodus opercularis*). *Behaviour* 90: 25–45

73 Francis RC and Barlow GW (1993) Social control of primary sex differentiation in the
    Midas cichled. *Proc Natl Acad Sci USA* 90: 10673–10675
74 Hostache G, Pascal M and Tessier C (1995) Influence de la température d'incubation sur
    le rapport mâle: femelle chez l'atipa, *Hoplosternum littorale* Hancock (1828). *Can J Zool*
    73: 1239–1246
75 Schultz RJ (1993) Genetic regulation of temperature-mediated sex ratios in the livebearing
    fish *Poeciliopsis lucida*. *Copeia* 4: 1148–1151
76 Baroiller JF, Chourrout D, Fostier A and Jalabert B (1995) Temperature and sex chromo-
    somes govern sex ratios of the mouthbrooding cichlid fish *Oreochromis niloticus*. *J Exp
    Zool* 273: 216–223
77 Baroiller JF, Fostier A, Cauty C, Rognon X and Jalabert B (1996) Significant effects of
    high temperatures on sex-ratio of progenies from *Oreochromis niloticus* with sibling sex-
    reversed males broodstock. In: Third International Symposium on Tilapia in Aquaculture,
    pp 333–343, Pullin RSV, Lazard J, Legendre M, Amon Kothias JB and Pauly D (eds),
    ICLARM Conf Proc 41. 11–16 November 1991, Abidjan, Côte d'Ivoire
78 Baroiller JF, Nakayama I, Foresti F and Chourrout D (1996) Sex determination studies in
    two species of teleost fish, *Oreochromis niloticus* and *Leporinus elongatus*. Zoological
    Studies 35(4): 279–285
79 Abucay JS, Mair GC, Skibinski DOF and Beardmore JA (1997) Environmental sex deter-
    mination: The effect of temperature and salinity on sex ratio in *Oreochromis niloticus* L.
    In: Proceedings of the Sixth Int Symp on Genetics in Aquaculture, 24–28 June 1997, Inter-
    national Association for Genetics in Aquaculture, Stirling, Scotland
80 Desprez D and Méland C (1998) Effect of ambient water temperature on sex determinism
    in the blue tilapia, *Oreochromis aureus*. *Aquaculture* 162: 79–84
81 Baroiller JF, Clota F and Geraz E (1996) Temperature sex determination in two tilapias
    species, *Oreochromis niloticus* and the red tilapia (Red Florida strain): effect of high or low
    temperatures. In: Proceeding of the Fifth International Symposium on the Reproductive
    Physiology of Fish, pp 158–160, Goetz F and Thomas P (eds), Austin, Texas, 2–8 July
    1995
82 Conover DO and Heins SW (1987) The environmental and genetic components of
    sex-ratio in *Menidia menidia* (Pisces: Atherinidae). *Copeia* 3: 732–743
83 Conover DO and Kynard BE (1982) Environmental sex determination: interaction of
    temperature and genotype in a fish. *Science* 213: 577–579
84 Conover DO and Heins SW (1987) Adaptive variation in environmental and genetic sex
    determination in a fish. *Nature* 326(6112): 496–498
85 Strüssmann CA, Calsina Cota JC, Phonlor G, Higuchi H and Takashima F (1996) Tem-
    perature effects on sex differentiation of two South American atherinids, Odontesthes
    argentinensis and *Patagonina hatcheri*. *Environmental Biology of Fishes* 47: 143–154
86 Strüssmann CA, Saito T, Usui M, Yamada H and Takashima F (1997) Thermal Thresholds
    and critical period of thermolabile sex determination in two Atherinid fishes, Odontesthes
    bonariensis and *Patagonina hatcheri*. *J Exp Zool* 278: 167–177
87 Strüssmann CA, Moriyama S, Hanke EF, Calsina Cota JC and Takashima F (1996) Evi-
    dence of thermolabile sex determination in pejerrey. *J Fish Biol* 48: 643–651
88 Blazquez M, Zanuy S, Carillo M and Piferrer F (1998) Effects of rearing temperature on
    sex differentiation in the European sea bass (*Dicentrarchus labrax* L.). *J Exp Zool* 281:
    207–216
89 Patino R, Davis KB, Schoore JE, Uguz C, Strüssmann CA, Parker NC et al (1996) Sex
    differentiation of channel catfish gonads: normal development and effects of temperature.
    J Exp Zool 276: 209–218
90 Yamamoto E (1995) Studies on sex-manipulation and production of cloned populations in
    Hirame Flounder, *Paralichthys olivaceus* (Temminck et Schlegel). *Bull Tottori Pref Fish
    Exp Stn* 34: 1–145
91 Conover DO and Fleisher MH (1986) Temperature sensitive period of sex determination
    in the atlantic silverside, *Menidia menidia*. *Can J Fish Aquat Sci* 43: 514–520
92 Baroiller JF and Jalabert B (1989) Contribution of research in reproductive physiology to
    the culture of tilapias. *Aquat Living Resour* 2: 105–116
93 Lagomarsino IV and Conover DO (1993) Variation in environmental and genotypic
    sex-determining mechanisms across a latitudinal gradient in the fish, *Menidia menidia*.
    Evolution 47: 487–494

94 Winge O and Ditlevsen E (1947) Color inheritance and sex-determination in *Lebistes.*
    Heredity 1: 65–83
95 Conover DO and De Mond SB (1991) Absence of temperature-dependent sex determina-
    tion in northern populations of two cyprinodontid fishes. *Can J Zool* 69: 530–533
96 Van den Hurk R and Lambert JGD (1982) Temperature and steroid effects on gonadal sex
    differentiation in rainbow trout. In: Proceedings of the International Symposium on the
    Reproductive Physiology of Fish, pp 69–72, Richter CJJ and Goos HJTh (eds), Pudoc,
    Wageningen, 2–6 August 1982
97 D'Cotta H, Guiguen Y, Govoroun MS, McMeel O and Baroiller JF (1999) Aromatase gene
    expression in temperature-induced gonadal sex differentiation of tilapia *Oreochromis nilo-
    ticus.* In Proceedings of the Sixth International Symposium on Reproductive Physiology
    of Fish, Bergen, Norway. 4–9 July 1999
98 Forest MG (1983) Role of androgens in fetal and pubertal development. *Horm Res* 18:
    69–83
99 Jeyasuria P and Place AR (1998) Embryonic brain-gonadal axis in temperature-dependent
    sex determination of reptiles: a role for P450 aromatase (CYP19). *J Exp Zool* 281: 428–
    449
100 Bogart MW (1987) Sex determination: a hypothesis based on steroid ratios. *J Theor Biol*
    128: 349–357
101 Shimada K (1998) Gene expression of steroidogenic enzymes in chicken embryonic
    gonads. *J Exp Zool* 281: 450–456
102 Kime DE (1993) "Classical" and "non-classical" reproductive steroids in fish. Reviews in
    Fish Biology and Fisheries 3: 160–180
103 Borg B (1994) Androgens in teleost fishes. *Comp Biochem Physiol* 109 C (3): 219–245
104 Spotila LD, Spotila JM and Hall SE (1998) Sequence and expression analysis of WT1 and
    Sox9 in the red-eared slider turtle, *Trachemys scripta. J Exp Zool* 281: 417–427
105 Kent J, Wheatley SC, Andrews JE, Sinclair AH and Koopman P (1996) A male-specific
    role for SOX9 in vertebrate sex determination. *Development* 122: 2813–2822
106 Clinton M (1998) Sex determination and gonadal development: a bird's eye view. *J Exp
    Zool* 281: 457–465
107 Eusebe DC, di Clemente N, Rey R, Pieau C, Vigier B, Josso N et al (1996) Cloning and
    expression of the chick anti-müllerian hormone gene. *J Biol Chem* 271: 4798–4804
108 Wibbels T, Cwan J and LeBoeuf R (1998) Temperature-dependant sex determination in the
    red-eared slider turtle, *Trachemys scripta. J Exp Zool* 281: 409–416
109 Baroiller JF and Clota F (1997) Interactions between temperature effects and genotype on
    Oreochromis niloticus sex determination. *J Exp Zool* 281: 507 (Abstract)
110 Baroiller JF (1996) Significant proportions of unexpected males in the majority of
    progenies from single pair matings with sibling sex-reversed males of *Oreochromis nilo-
    ticus.* In: Third International Symposium on Tilapia in Aquaculture, pp 319–327, Pullin
    RSV, Lazard J, Legendre M, Amon Kothias JB and Pauly D (eds), ICLARM Conf Proc 41.
    11–16 November 1991, Abidjan, Côte d'Ivoire
111 Bournon C, Houillon C and Pieau C (1990) Temperature sex-reversal in Amphibians an
    Reptiles. *Int J Dev Biol* 34: 81–92

Index